The Essentials Series for Business-Driven Software Development

Lean-Agile Pocket Guide for **Scrum Teams**
second edition

Alan Shalloway and James R. Trott

Lean-Agile Pocket Guide for Scrum Teams

Alan Shalloway and James R. Trott

Net Objectives Press, a division of Net Objectives
1037 NE 65th Street Suite 362
Seattle WA 98115-6655
Toll Free: 1-888-LEAN-244

Find us on the Web at: *www.netobjectives.com*

To report errors, please send a note to *info@netobjectives.com*

The Lean-Agile Pocket Guide for Scrum Teams is a condensed version of materials offered by Net Objectives. It is intended to support those who have already received training. It is appropriate for any role on the Scrum team.

Copyright © 2014 Net Objectives, Inc. All Rights Reserved.

Written and edited by Alan Shalloway and James R. Trott.

Net Objectives and the Net Objectives logo are registered trademark of Net Objectives, Inc.

Notice of Rights

No part of this publication may be reproduced, or stored in a retrieval system or transmitted in any form or by any means, electronic, mechanical, photocopying, recording, or otherwise without the written consent of Net Objectives, Inc.

Notice of Liabilities

The information in this book is distributed on an "As Is" basis without warranty. While every precaution has been taken in the preparation of this book, neither the authors nor Net Objectives shall have any liability to any person or entity with respect to any loss or damage caused or alleged to be caused directly or indirectly by the instructions contained in this book or by the computer or hardware products described in it.

ISBN 978-0-9713630-2-1

10 9 8 7 6 5 4 3 2
Printed in the United States of America

Second Edition

Contents

Preface ... ix
- Continuous Improvement ... x
- Acknowledgements ... x

Introduction ... 1

Part I: Essential Concepts ... 3

Essential Competencies of Scrum ... 3
The Four Essential Practices ... 5
Choosing an Agile Approach ... 5
- Organizational and Team Capacity ... 5
- Potential Risks ... 6
- Potential Rewards ... 6
- Starting an Agile Team Checklist ... 7
- Issues and Considerations ... 8

The Drivers of Lean-Agile Thinking ... 9
- Basic Motivations ... 9
- The Reasons for Going Agile ... 10
- Agile Follows a Regular Life-Cycle ... 12
- Lean Principles and Agile Practices ... 12
- Agile Involves a Commitment to Continuous Improvement ... 14
- Agile Involves Oversight of the Flow of Value ... 15

Principles and Practices of Lean-Agile ... 16
- Foundations ... 16
- Principles ... 17
- Practices ... 18

Essential Skills for Agile Developers ... 19

Part II: Roles and the Team ... 21

Roles by Level ... 22

- The Portfolio-Level 22
- The Management Level (application areas) 22
- The Front-Line Level 23
- A Note about the ADM and TDM Roles 24

Business Product Owner 27
Scrum Master 29
Business Analyst (Business SME) 32
Developer 34
QA / Tester 36
Release Manager 38
Database Administrator 40
Database Developer 41
A Team of Peers: Roles in Lean-Agile 42
Swarming and Teamlets 44

Part III: From Concept to Consumption   45

Overview: Business Discovery and Business Delivery 45
- Phases of Business Discovery and Business Delivery 46
- Progressive Unfolding 48

Activities of Business Discovery: Priority, Planning, and Staging .. 49
- Checklists 49
- Prerequisites 49
- Activities 49
- About Estimation and Prioritization 50

About Planning and Staging 51
- Checklists 51
- Prerequisites 52
- Activities 52
- The Thinking Activities of Planning and Staging 52
- Analysis: What we have to provide the Business 54
- Selecting: The minimum required to realize value 55
- Adjusting: Releases and the Business Case 56
- Selecting: Elevations 57

Ordering: When we need to provide it 58
Issues and Considerations 59
Iterations 60
Checklists 60
"Iteration 0" 61
Checklists 61
Prerequisites 61
Activities 61
Iteration Planning 62
Checklist 62
Prerequisites 62
Activities 62
Add, Revise, Remove Requirements 63
Issues and Considerations 63
Iteration Execution 64
Checklists 64
Prerequisites 64
Activities 64
Starting Stories 64
Selecting Tasks 65
Updating the Story Board 65
Completing Stories 66
The Daily Stand-Up (Daily Scrum) 67
Checklists and Ground Rules 67
Prerequisites 67
Activities 67
Issues and Considerations 69
Product Demonstration 70
Ground Rules 70
Prerequisite 70
Agenda 70
Issues and Considerations 71
Elements: Vision for the Product 72
Prerequisites 72

v

Activities	72
Elements: Features	**74**
Elements: Stories	**75**
Elements: Tasks	**77**
Estimation in Agile: A Conversation	**79**
Estimating by points	79
Estimating work to be done	79
Estimating team capacity	80
Issues and Considerations	80
Team Estimation Game	81

Part IV: Quality   83

Continuous Improvement Before, During, and After	**83**
Retrospection After Every Iteration	**84**
Prerequisite	84
Activities	85
Facilitator Notes	86
Issues and Concerns	87
Impediments to Progress and Quality	**88**
Addressing Impediments	88
Issues and Considerations	89
Examples of Impediments	89
Quality: Documentation, Standards	**92**
The Lean-Agile Approach to Standards	93
Patterns and Code Quality	93
Attributes of Code Quality	94
Quality: Testing	**95**
Acceptance Testing	95
Automated Testing	96
Continuous Integration	**100**
Automated, scripted builds and testing	101
Using a dedicated CI server / build farm	101
Triggering CI server on check-in	102

Build as frequently as possible 102
Build the entire system 102
Run automated test suites 103
Notify team(s) when build breaks or a test fails 103
Beyond Basic CI 104
Communicating Information 108
The Scrum Team Room 108
Information Radiators & the Team Project Board 108
Considerations for Non-Co-Location 110

Part V: Checklists for Lean-Agile   111

Starting an Agile Team Checklist 111
Business Discovery: Priority Checklist 113
Business Discovery: Planning Checklist 113
Business Discovery: Staging Checklist 114
Business Discovery: Ready-to-Pull Checklist 114
Business Delivery: Iteration 0 Checklist 115
Business Delivery: Iteration Planning Checklist 117
Business Delivery: Iteration Execution 119
Business Delivery: Product Demonstration 120
Business Delivery: Release Checklist 120
Business Delivery: Iteration Implementation 121
Daily Stand-Up 122
Schedule: Various Scrums 124
Schedule: Iteration Planning 125
First Day of Iteration Planning: Morning 125
First Day of Iteration Planning: Afternoon 126
Second Day of Iteration Planning: Morning 126

Part VI: Resources   127

A Lean-Agile Glossary 127
Recommended Resources 144
Net Objectives Resources Page 144

Net Objectives Scaled Agile Framework Resources 144
Net Objectives Training for Teams and Roles 144
Net Objectives Agile Design and Patterns Resources............................144
Essential Skills for the Agile Developer..144
Acceptance Test-Driven Development Resources................................145
Recommended Web Sites..145
Recommended Books..145

Books from Net Objectives 148

Ordering and Licensing Information 152
License (Private Label) this Pocket Guide for Your Company 152

Index 153

Preface

The Lean-Agile Pocket Guide for Scrum Teams collects in one spot the good practices we have learned and observed as we have trained thousands of teams in Lean and Agile software development, including Scrum.

As with all "pocket guides," this book is most helpful as a supplement to training, such as the courses described at *www.netobjectives.com/training/scrum-and-kanban*. Our objective is to offer guidance through terse descriptions and checklists and visuals rather than trying to instruct you in essential concepts. It is organized to help you find what you need when you need it, following the normal life of a project.

This guide is scoped to support the Scrum team, the business analysts and developers and testers and Scrum Master and project manager who work together to create a product. It covers everything that they will be involved in as they do their work. Certainly, there is more to Lean-Agile than this, both upstream (in product planning) and downstream (in release and support) as well as managing the value stream. We chose to focus on the team to keep the guide at a reasonable and useful size.

LEAN-AGILE is not cut-and-dried

Lean-Agile covers many principles and practices. There are several flavors of Agile methods in the marketplace, Scrum being the most popular at the moment. To make this guide the most widely useful, we chose to use the following terminology. The techniques are similar even if the words are different. Substitute as you see fit.

- **Iteration.** The time-boxed period during which teams focus on producing demonstrable product. Scrum calls this a "sprint."
- **Scrum Master.** The role responsible for the process and health of the team. We chose to use this Scrum term because it is so widely used, even by teams not doing Scrum. This may also be known as the team facilitator or even the project manager.
- **Product Vision.** The short statement of the vision that is driving the product work, expressed in business and customer value terms. This is often known as the project charter.
- **Populate.** When someone adds an item of work to the backlog of work to be done by the team, we say they "populate the backlog" or "place the item onto the backlog." Some methods call this "loading the backlog" but this is not widely understood.

Continuous Improvement

Quality improvement is central to Lean and Agile. This is true in this pocket guide as well. As we learn more, the guide will improve and expand. We will continue to improve and expand. As you find areas for improvement or correction, please send us an e-mail. Our contact information is at *www.netobjectives.com*.

Acknowledgements

Many people have contributed significantly to this guide. In particular, we wish to acknowledge Alan Chedalawada, Guy Beaver, and our colleagues at Net Objectives who are some of the best thinkers we have found in this area. Andrea Bain has been so helpful with the graphics.

We wish to acknowledge Curt Hibbs who has been tireless in proofing and contributing valuable ideas. Kate Bogh helped us develop a professional glossary of Lean and Agile terms. Alex Sidey has been an inspiration and mentor in writing and quality.

The concepts in this guide build on shoulders of giants in the industry. They represent the practical combination of Lean and Agile / Scrum principles and practices applied to software development. Some of the concepts and methods present have been created by Net Objectives (especially in the area of release planning) and some represent the guidance of others. Space does not allow us to acknowledge everyone, but here are some of the most important in this field:

Christopher Alexander	Jane Cleland-Huang
Mark Denne	Daniel Jones
David Mann	Mary and Tom Poppendieck
Don Reinertsen	Peter Scholtes
Jeff Sutherland	David Womack

Introduction

This pocket guide is designed to reinforce the Lean-Agile thinking that you, as a member of a Scrum team are developing as you use Scrum to develop software products. To get the most use of this tool, you and your team should have already had a training course, such as *Implementing Lean-Agile for Your Team* by Net Objectives.

Scrum is much more a way of thinking about the process of software product development, than it is a particular set of practices. This guide offers a brief review of the basic drivers and motivations of Scrum and the principles and practices that form the Lean-Agile way of thinking, which underpins Scrum. You and your team are responsible for the process you use to create products. Before you can improve, you must know how to think in this new paradigm of Lean-Agile. It will help you avoid much wasted time and effort.

> **An activity on a project is value-added if it transforms the deliverables of the project in such a way that the customer recognizes the transformation and is willing to pay for it. (Mascetelli, 2002)**

In Lean-Agile software product development, your goal is to meet your customer's value and quality objectives, to deliver as much value to the customer as they can consume, as quickly as possible, in the most efficient manner possible, in a sustainable way.

Following this brief description of the principles and practices, you will find sections on the Roles of Scrum, planning, analysis, and estimation. In Scrum, a review of quality and continuous process improvement techniques, and some notes on special cases.

This pocket guide is designed for the Scrum team. Members of the Scrum team will find most of what is in this to be useful as they do their daily work. There are many special roles in Lean-Agile: the Business Product Owner, the business owner, the Scrum Master, management. These roles interact with the Scrum team but have their own set of practices and techniques they use. They will find some of this pocket guide useful to their daily work, but that is not the primary audience.

I Essential Concepts

Part I: Essential Concepts

Lean-Agile involves a number of core, essential concepts that everyone must learn. This part describes these concepts.

Essential Competencies of Scrum

Net Objectives offers a certification process for Scrum team members. The exam covers the following essential competencies. These are covered later in this pocket guide and in training.

Business Case for Scrum

Always drive from Business value

The business drivers for Scrum

How Scrum helps mitigate risk in software development projects

An activity on a project is "value-added" if it transforms the deliverables of the project in such a way that the customer recognizes the transformation and is willing to pay for it. Other activities represent waste and should be reduced or eliminated.

Test first, Acceptance Test-Driven Development, design patterns, and the sustainable development practice

A process is an agreement by a team about how they will work together. It may be changed by the team but otherwise must be followed.

Understand the problems of requirements.

Essential Aspects of Agile

The Daily Stand-up

The Goal: Incremental delivery of value through iterative development according to the capability of the team.

The "Minimal Business Increment"

Features: Sizing, language and terminology

Good features have these characteristics: Independent, Negotiable, Valuable, Estimatable, Small, Testable (INVEST)

The essential elements of a story: Size, Description, Done Criteria, Value

Tasks are written by and for the team. Elements of a task: Description, Owner, Estimate

Essential Concepts

Roles in Scrum

The **Business Product Owner** forms the product vision, improves ROI, manages customer and stakeholder expectations, road-mapping and release planning, prioritizes the product backlog, provides clear, testable requirements to the team, collaborates with both customer and team to ensure goals are met, and accepts product at the end of each iteration.

The **Scrum Master** shepherds the team, creates a trustful environment, facilitates team meetings, asks the difficult questions, removes impediments, makes issues and problems visible, keeps the process moving forward, and socializes Scrum within the greater organization.

The **Team** estimates size of backlog items, makes design and implementation decisions, commits to increments of deliverable software, and delivers it. The team tracks its own progress, is autonomous and self-organizing, and is accountable for delivering as promised.

A **Swarm** is a temporary group of team members that works together on one story to bring it to completion.

Quality Assurance focuses more on discovering why defects are occurring than in finding defects.

Good Scrum Practices

Iteration 0 prepares the team for the release and what they need to know to start work quickly.

How to prioritize / order stories in an iteration.

The best designs emerge from self-organizing teams using iterative development.

Story points are arbitrary units that teams use to represent levels of effort and do not necessarily correspond to hours or days.

Scrum teams should be continually improving their processes, which they own. Management helps teams think through their processes.

A "right-sized" story can be completed within an iteration. Usually, stories are sized to be completed within 1/3 to 1/2 of the iteration.

Co-located teams involve being near each other and intentionally working together in a close collaboration to develop product.

Iteration reviews and retrospections are for the team and the customer to promote continuous improvement.

Before stories can be started, they must have validation or acceptance criteria.

The team is protected during the iteration.

The Four Essential Practices

Our experience shows that Scrum teams that incorporate these four practices into their work always improve their performance.

- Manage work in process to some level anyway
- Use explicit policies
- Have work be visible
- Include management

Choosing an Agile Approach

The primary reason for choosing an Agile approach is that it will create a significant return on the investment. This is always a judgment call. Here are some of the important factors to consider in this decision.

Organizational and Team Capacity

Consideration	Think about...
Does the organization have the capacity to make the change?	Close contact with customers so they can prioritize features
	Committing to incremental delivery of value through iterative development
	A commitment to greater customer involvement and prioritizing for customer value
	Personnel policies that encourage team-based performance
Do teams have the capacity to adopt new ways of working?	Commit to completing software every iteration
	Minimize work in process during an iteration
	Minimize interruptions and multi-tasking
	Access to just-in-time training and learning options
	Co-location as much as possible
	Access to the tools and environment in time for the first iteration
	A trained Scrum Master is available and focused on the team

Potential Risks

Consideration	Think about...
During the transition to Agile, risks could be costly and must be mitigated	The impact of errors while learning how to prioritize work
	The challenge of creating stories small enough to be completed within an iteration
	Large projects involving multiple teams can be difficult to coordinate
	Learning to write changeable software and spending the time to refactor as more is learned
	Methods of allocating resources (and accounting rules) may not allow one project per person
	Costs for conducting much more regression testing early on

Potential Rewards

Consideration	Think about...
Most teams realize some benefits right away	Improved team coordination
	Improved morale from more frequent quick wins and from catching mistakes earlier
	Applying team knowledge to local conditions for more relevant processes
The organization realizes a number of savings over the long-term	Improved code quality and lower technical debt, which allows for easier change
	Regression testing cost will be lower over the long term as code is perfected
	Better use of resources, less wasted effort
	Less unnecessary and unwanted code that, nevertheless, must be maintained
Customer satisfaction improves with more involvement	Quicker access to features that are valuable to do work "now"
	Less "bloatware": features that are not wanted, rarely used
	Stopping projects early and being able to re-deploy teams on to other projects when the customer is satisfied with what has been delivered

Starting an Agile Team Checklist

✓	Activity	Description
	Select the right people	Business Product Owner. *This is crucial.* The ADM and TDM The Scrum Master Candidates for the team
	Select the right project	Select candidate projects that have a good likelihood for visibility and success. Select projects that will help you learn and that address dynamics of the organization.
	Train the people	Allow enough time to train the team with the skills they require. Allow enough time to give developers specialized training in areas such as • Test-Driven Development • Object-orientation and design patterns • Automated acceptance testing
	Build the physical environment	Establish and/or build out the physical environment for the team. • The Scrum Team room • Visual controls and team project board Decide how to handle teams that are not co-located.
	Tools	Acquire, build-out, and configure tools and environments for the team • Agile life-cycle management tool • Source control management • Continuous integration server • Test frameworks
	Plan for assessment	Develop a plan for assessing maturity of Scrum Teams. Develop a plan for assessing Agile project success.
	Transition plan	Develop a plan for making the transition. Gather Scrum Masters into a community of practice. Gather Business Product Owners into a community of practice.

Essential Concepts

Issues and Considerations

Topic	Discussion
Getting team buy-in	Candidates for an Agile team do not always buy into this new process. They have to complete the training and get their questions answered. Then, if they do not buy in, let them do something else. Scrum team members must all agree with this approach or it will likely fail.
Lack of tools	The hardware and tools must available from the start, right at the first iteration.
Management buy-in	Management has to understand the process and agree not do things that create impediments.

Please see "Net Objectives Training for Teams and Roles" on page 144.

The Drivers of Lean-Agile Thinking

This section summarizes some of the basic drivers of Lean-Agile thinking. The Net Objectives website offers an extensive collection of resources about Lean-Agile thinking applied to Scrum. See *www.netobjectives.com/resources/lean-scrum*.

Basic Motivations

Agile software development implements the principles of Lean software development at the local team level. Lean thinking extends the benefit across the enterprise. Together, this creates a framework that:

- Enables the team to discover what the customer needs and to deliver value quickly and repeatedly
- Enables the whole team to remove impediments that are in the way of delivering value quickly
- Encourages continuous refinement of the framework to increase the accuracy of what is delivered
- Optimizes the flow of value delivery across the whole enterprise

The primary features of Lean-Agile include:

- Building software in iterations
- A product backlog that is prioritized to deliver the most business value
- Using cross-functional teams where people use their skills to get work done, using a facilitator
- High-bandwidth communication as much as possible
- Daily Stand-up meetings for the whole team
- Simple visual controls at the work site to report progress and issues
- Risk mitigation: address issues, impediments, and dependencies early
- Continuous process improvement and best practices

Essential Concepts

The Reasons for Going Agile

Need	How Lean-Agile Addresses the Need
Add value to the Business quickly	The Business prioritizes work to increase revenue/benefit, to increase market position and to increase feedback (and reduce risk).
	The product team is a partner in the enterprise value stream.
Gain clarity on customer needs	Work on aspects of features that customers are *clear* about.
	Use short iterations to increase feedback and grow knowledge.
	Minimize speculation; let the design emerge with experience over time.
Manage projects better	Metrics are reported using simple visual controls (also known as "information radiators"), with an intense focus on value creation, current work, and reducing impediments.
	Radiators are managed locally and are quite visible.
	Work is planned and re-planned constantly; however, the planning horizon is short. Perhaps 15% of the team's time is spent on planning: releases, before each iteration, adjusting during the Daily Stand-up.
	Testing is central to development. The goals of testing are to ensure high-quality code at each iteration, finding bugs, discovering root-cause issues that lead to defects, and creating acceptance tests for every feature and every story.
	Short deadlines and close contact with the customer ensure a constant sense of urgency without panic. The Business Product Owner is rarely surprised.
Motivate teams	Short feedback cycles and close contact with the customer results in more quick wins, which motivates the team. Catch mistakes early when there is still time to effect change.
Product-centered development	Let the design emerge as the team gains experience with what is needed. This reduces the tendency to over-design.
	The deep use of patterns and agreed-upon standards ensures that what is produced can be extended appropriately with a minimum of complexity.

Agile Is Built on Twelve Essential Principles

Principle	Description
Satisfy the customer	The highest priority is to satisfy the customer through early and continuous delivery of valuable software
Welcome change	Welcome changing requirements, even late in development. Agile processes harness change for the customer's competitive advantage.
Deliver frequently	Deliver working software frequently, according to how quickly the customer can consume it.
Whole team, daily	The Business and developers must work together daily throughout the project
Motivated people	Build projects around motivated individuals. Give them the environment and support they need, and trust them to get the job done.
High bandwidth	Face-to-face conversation is the most efficient and effective method of conveying information to and within a development team.
Measure of success	Working software is the primary measure of progress.
Sustainable	The sponsors, developers, and users should be able to maintain a constant pace indefinitely.
Technical excellence	Continuous attention to technical excellence and good design enhances agility.
Simplicity	The art of doing just enough and no more than is necessary.
Emergent design	The best architectures, requirements, and designs emerge from self-organizing teams.
Continuous improvement	At regular intervals, the team reflects on how to become more effective and then adjusts its processes and behaviors to do this.

Agile Follows a Regular Life-Cycle

The life-cycle of Scrum starts with the product vision, runs through some form of product portfolio management to gather overall requirements and prioritize and then iteratively develops and releases product so that the Business can realize value quickly, all the while learning more about what the customer needs.

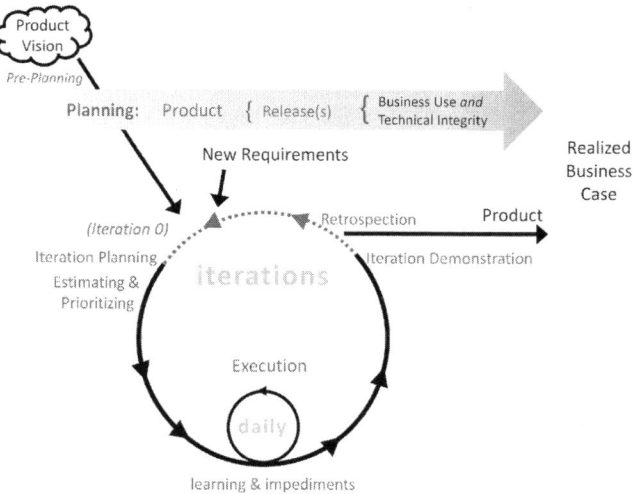

Lean Principles and Agile Practices

Lean has two primary emphases: Achieve fast-flexible-flow and discover what is needed. These emphases involve seven key principles (Poppendieck and Poppendieck 2003). The practices of Agile map to these principles. These principles work together to remove delays and to minimize the amount of work in process. When work-in-progress (WIP) is high, the team will thrash and there is a risk of greater waste if directions have to change.

Principles and Practices: Achieve Fast-Flexible-Flow

Lean Principle	Agile Practices	
Optimize the whole	Product vision	Product backlog
	Co-location	Business value burn-up
	Release planning	Business value
	Story point sizing	Feature/Story sequencing
	Line of sight on release and iteration goal	
Eliminate waste	Process improvement	Kaizen
	After Action Review	Root Cause Analysis
Build quality in working software	Validation-centric	Acceptance TDD
	Story/Task doneness	Continuous integration
	Integrate development team and QA	
Deliver fast	Establish a regular iteration cadence (1-4 weeks)	
	Iteration planning	Backlog management
	Iteration commitment	Value-creation metrics
	Iteration review	

Principles and Practices: Discover what is needed

Lean Principle	Agile Practices	
Defer commitment	Pull	
	Just-In-Time Stories	
	Story Unfolding (decomposition)	
Create knowledge	Information metrics	Dashboards
	Daily accountability	After Action Review
	Iteration retrospection	Dependency tracking
	Impediment tracking	Velocity tracking
	Community of practice	Collaboration
Respect people	Leader standard work	
	Boundary condition visibility	

Essential Concepts

Agile Involves a Commitment to Continuous Improvement

Processes are never perfect nor are they universally applicable. Processes support the team and not the other way around. A process describes how a team of people work together. They agree together to follow the process and, when they discover that something is not working, to stop and change it and then follow it again.

The team is responsible for its processes, improving based on its knowledge of local conditions.

Managers facilitate a spirit of continuously improving processes: empowering people to improve and expecting that it is done. They hold teams accountable for results. Managers guide teams by asking good, thought-provoking questions to get people to think clearly about their problems.

Process improvement activities need not take long. But they do require a commitment and *discipline* to stop, adjust, and then go on. This builds the team's confidence in itself, its ability to solve problems, and its ownership of its own processes.

> **Tip.** Continuous improvement is not easy. Often, teams over-focus on the need to deliver on the iteration goal and don't see process problems. The Scrum Master and managers serve the team by calling attention to process problems and facilitating After Action Reviews.

Agile Involves Oversight of the Flow of Value

Oversight concerns the control of time, money, and quality relative to the flow of value. It is the responsibility of value-stream management, the Business Product Owner, the Scrum Master and the team. Short planning horizons make this oversight easier and more effective.

At this level...	Oversight involves these activities
Overall product management	Business Product Owner works with the Business to prioritize projects, capabilities, and features within the planning horizon. Sets goals for initiatives and releases.
	Business Product Owner advocates for the product to ensure clear communication and progress.
Project	Business Product Owner creates the product charter, explains it to the team. The product charter is displayed in the team's work area.
	At releases and iterations, status is reviewed against the product charter.
	Daily review of business value burn-up (feature completion) and impediment burn-down.
	Create the project board for the team.
Iteration	Daily Stand-ups cover progress relative to priorities.
	Visual controls are updated daily by the team and the Scrum Master so that status is always visible to the team and to management.
	The Business Product Owner and other management make a practice of "going to the Gemba" (where the team is working) to get a sense of what is going on rather than simply relying on reports.
	Local teams assume responsibility for governing themselves in terms of process and process improvement and standards and practices to use.
	Continuous builds assure near-perfect code is always maintained.

Principles and Practices of Lean-Agile

Lean-Agile is built on a foundation of respect, value, and system thinking. The Lean-Agile organization embraces a set of principles and practices that create a culture fostering successful projects.

Foundations

Respect People. People hold the knowledge about processes, problems, and improvements. Teams use their local knowledge, guided by enterprise standards, to create processes to deliver product. Each team member must deliver on his or her own commitments and depends on others to do the same. The Business Product Owner trusts the team to deliver on its commitments.

Continuous Process Improvement. Processes exist to serve the people, to help them get their work done. No process is perfect; when problems arise, the team is responsible to *stop*, *change the process*, and *start again*. Without affixing blame to any one person.

Focus on Value. Team decisions are based on what delivers business value quickly. The Business prioritizes what features are necessary to satisfy needs of the Business and its customers.

Remove Impediments. The team is responsible for identifying and, as much as possible, removing anything that impedes their ability to deliver value. If it is beyond their scope of control, then management assumes responsibility. Impediments are tracked.

Principles

Optimize the Whole. Focus on the entire portfolio of products that are in the "value stream" for the enterprise. Do what is needed to improve the overall flow of the most important value. At the project level, business analysts, developers, and testers work together to make high-quality code. Bugs are not tolerated.

Eliminate Waste. Eliminate anything in the product development process that creates waste, anything that does not add business value or customer value, that has been started but has not been put into production, that delays development, or that is more than what the customer requires (or goes beyond delighting the customer).

Build Quality In. Quality is *built into* the product, not tested in. It is part of design. Patterns inform design choices that create sustainable, less complex, more flexible code. Quality assurance and testing exist less to catch bugs than to identify causes of bugs so that processes and systems can be corrected. Code development is test-driven.

Create Knowledge. Each development project, environment, and team is unique. Each iteration offers a learning opportunity. Knowledge that is not transferred or is not embedded in processes is lost. Be intentional about discovering lessons and facilitating knowledge exchange in a safe, blame-free way. Learn what customers know and let the requirements emerge based on what they learn while doing.

Use Visual Controls. Progress charts, product vision, rules, and instructions, are visible everywhere. Teams use regular, common meeting rooms for high bandwidth communication.

Test Early, Test Often. Validate that the team understands what the Business needs, that the product produced is what was required, and that the product is of sufficient quality. Move testing as early as possible into the development process. No feature or story is created without a corresponding acceptance test.

Foster a Team of Peers. Communication between roles must be unrestricted and transparent, based on mutual respect and personal safety. All roles are responsible for product quality, must act as customer advocates, and must understand the business problem they are trying to solve.

Practices

Use Agile / Scrum Development. Teams should follow the Scrum framework to product development. Ideally, the whole team receives training in Scrum before they begin so that everyone is on the same page. Every team has a Scrum Master assigned to them to facilitate their work.

Deliver Incremental Value. Small, quality deliverables show progress, promote learning, and improve team morale; help detect risks, bugs, and missing requirements early; and forces key decisions when feedback can still make a difference. Keys to frequent delivery include: keep the scope of an activity small, work on deliverables in a "just-in-time" manner; leaving options open by eliminating premature decisions; use continuous build and test. Plan, execute, and measure progress and team velocity based on the delivery of increasing value.

Foster Open Communication. Historically, many organizations and projects have operated purely on a need-to-know basis, which frequently leads to misunderstandings and impairs the ability of a team to deliver a successful solution. Lean-Agile requires open and honest communications, both within the team and with key stakeholders. A free-flow of information not only reduces the chances of misunderstandings and wasted effort, but also ensures that all team members can contribute to reducing uncertainties surrounding the project.

Stay Agile, Adapt to Change. The more an organization seeks to maximize the business impact of a technology investment, the more they venture into new territories. This is inherently subject to change as more is learned about what is needed. It is impossible to isolate a project from these changes. All core roles must remain available throughout a project so that they can contribute to decisions arising from these changes. The team works together to address these issues based on the knowledge each member has. The contribution of all team roles to decision-making ensures that matters can be explored and reviewed from all critical perspectives.

Practice Good Citizenship. Team members are stewards of corporate, project, and computing resources. This attitude manifests itself in many ways: conducting the project in an efficient manner; optimizing the use of system resources; reusing existing resources. Team members respect each other's time and skills and seek to live up to commitments made to each other. Each team member is responsible for an environment that is clean and accessible, agreeing on a discipline that helps them sustain the environment.

Essential Skills for Agile Developers

When an organization is transitioning to Agile, developers often raise many concerns, especially how they are supposed to write code in an environment in which requirements are changing on them and there is little or no architecture. Addressing those concerns involves gaining knowledge of Agile development and acquiring certain skills that are essential to thrive in an Agile environment.

Each of the following skills is easy to learn and yet provides huge improvements to the developer's code. The resources section describes a number of on-line resources to help you acquire and improve them.

Skill	Description
Minimize complexity and rework	This is more of an attitude than a specific skill set. It provides an alternative to the common extremes of hacking (causing much rework) and over-design (causing greater complexity than needed).
Keep the system in a running state	Finding errors is the developer's biggest time waster. When several changes occur and then something goes wrong, it is hard to know what change caused the error. By making incremental changes to detect errors immediately, this saves a lot of wasted time searching for errors.
Programming by intention	This old programming technique (made popular in eXtreme Programming) is easy to do and yet provides decoupling, cohesion, encapsulation and clarity. If there is one practice you can (and should) easily pick up, this is it.
Separate use from construction	This eliminates the situation in which integrating new code into established systems exceeds the cost of writing the code in the first place. It forces abstraction and encapsulation onto the developer. It provides discipline for hiding the specific implementation you are using.
Commonality and variability analysis	A straightforward technique to identify the concepts in your problem domain. While developers are capable of thinking in terms of abstractions and implementations, few do. This provides a model to further the proper use of object-oriented modeling.

Essential Concepts

Skill	Description
Encapsulate that!	Encapsulation lives at all levels. We are often unsure of how we should implement something. We often make a choice which proves bad and requires rework or we overbuild increasing complexity. If we encapsulate our code and designs we can more efficiently make needed changes when we understand what is really needed.
Define tests up-front	Define tests up-front even if you do not write them up-front. It takes no work, increases understanding, and prevents many errors.
Shalloway's Law and Shalloway's Principle	Shalloway's Law states that if you have to find more than one thing when a change is made, you will miss something. Shalloway's Principle says to avoid situations where Shalloway's Law applies: Use proper use of object-orientation and encapsulation so that you never have to find more than one thing when a change is required.
When and how to use inheritance	Inheritance has a long heritage of improper use. The seminal book *Design Patterns: Elements of Reusable Object-Oriented Software* provided a better way to use it. This chapter gives a quick overview to their approach – one that results in more extensible systems.
Refactor to the open-closed	While refactoring can mean fixing code, it also is used to improve designs that were good when they were written but decayed because of new requirements. This technique improves the design of quality code as more knowledge about the system's needs become available.
Needs vs. capabilities in interfaces	Interfaces provide a different method of design. Instead of designing from the perspective of what we have, we should be designing from the perspective of what is needed. This follows the concepts of design patterns as well as several object-oriented design principles.
Continuous integration	Continuous integration is a low cost method of detecting errors quickly. It is essential if you are writing tests up front.

Part II: Roles and the Team

In Lean-Agile, the delivery of value is the effort of people all across the value stream: executive, management, and front line teams. It requires people from both the Business and technology working together to discover and deliver value to customers so that they can use it and realize value from it.

In this scheme, all of the roles are needed and people may play more than role. We like to show this as a pyramid involving three levels and a partnership of two halves: Business and technology.

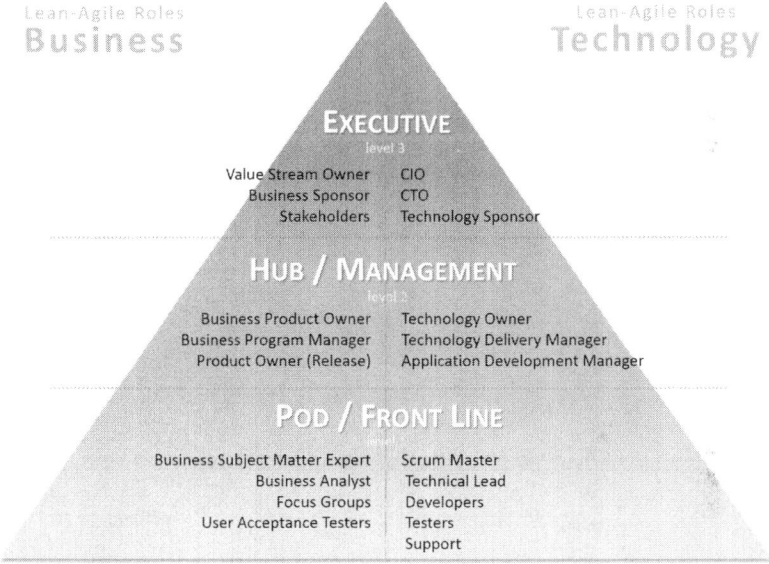

This section provides the following:

- **Summary of roles by level.** A summary of the work of the three levels.
- **Standard work for the roles.** The responsibilities and characteristics of some of the roles This can serve as a starting point for defining how this role should look in your own organization.
- **Team of Peers.** The special notion in Scrum that teams are composed of "peers" working together and what that means. Sometimes, this requires using "swarms" and "teamlets."

Roles by Level

The Portfolio-Level

The Portfolio includes both projects and additional "reserved" capacity for enhancements, production support, and maintenance requests. Roles involved in the portfolio include the Value Stream Owners and the sponsors. Portfolio management is responsible for priority, the order of work, and deciding what is active. It makes these decisions based on total capacity Pods. It also assigns Business Product Owners.

Role	Focus of the Role
Value Stream Owner	Continual highest Business value
	Optimal cycle time from idea to realization
Business Sponsor	Realize Business value and ROI
	Provide final approvals and resources/funding
Technology Sponsor	Technology and process to incrementally realize Business value with quality solutions
Stakeholders	Business standards for the organization

The Management Level (application areas)

The Management Level (sometimes called the "Hub") is the center of responsibility that is focused on an application area. Work is comprised of releases, enhancements, production support, and maintenance requests.

Role	Focus of the Role
Business Product Owner / Product Management	Incremental Business value (highest ROI)
Business Project Manager	Visibility and progress of portfolio Book of Work and/or program backlog
Technology Delivery Manager	Continual incremental delivery of quality solution(s)
Application Development Manager / Technical Lead	Application / System Integrity

Roles by Level

Role	Focus of the Role
Information Risk Manager	Provides guidance on compliance with required application controls
Infrastructure PM	Responsible for establishing and maintaining project artefacts and execution for Infrastructure
IT Project Manager	Pool resource that supports skills for the project administrative support required by the life cycle
Infrastructure Engineer	Responsible for executing the design and implementation of infrastructure

The Front-Line Level

A Front-Line level (sometimes called the "Pod") is the "whole team" including developers, testers, analysts and SMEs. Work is composed of the stories and tasks for a specific release, enhancements, production support, and maintenance requests.

Pod members are responsible to produce and implement a Business value increment, for quality assurance, and continuous incremental improvement.

Skills needed in the Pod include analysis, design, code, build, deploy, testing, acceptance, validation, and implementation. Extended skills needed include: Architecture and design, build and release management (environments), and subject matter/domain expertise (technical, Business, and customer).

Role	Focus of the Role
Product Owner	Acceptance, validation, adoption: MBI and features
Scrum Master	Visibility and continuous incremental improvement
IT Project Manager	Program / project / process life-cycle management
Subject Matter Expert	Acceptance criteria
	Validation and implementation of business value increment
	Contribute to requirements
Whole Team	Incremental value delivery, quality, and continuous improvement

23

Role	Focus of the Role
Analyst	Synthesizing/refining requirements
Technical Architect	System architecture and design
QA Tester	Represent testing discipline within the team
	Develop and execute test cases
	Coordinating QA activity
Production Support	Provide operational and production input.
	Prepare and execute transition to production

A Note about the ADM and TDM Roles

The role names Technology Delivery Manager (TDM) and Application Development Manager (ADM) might not be familiar but the responsibilities should be. This section offers a compare and contrast of these roles so you can identify who plays these roles in your organization.

Application Development Manager

The Application Development Manager (ADM) is the person who has accepted responsibility for the holistic integrity and functionality of the application. The ADM is responsible for risks associated with ***architecture and quality***. This includes:

- Think and acts at the system level
- Build the system holistically
- Oversee implementation and test integrations across all subsystems
- Coordinate work across multiple teams to ensure integrity
- Manage conflicts, collaboration and dependencies
- Ensure the iterative engineering practices and standards to produce quality solutions incrementally

The ADM role includes but goes beyond "System Architect" or Delivery Manager. Architects can specify design reviews and architectural standards; however, they usually don't get to determine technologies, languages, coding standards. The ADM does all of this with a focus on overall integrity of the application.

Technology Delivery Manager

The Technology Delivery Manager (TDM) is the person who has accepted responsibility to maintain the technical integrity of the system. The TDM is responsible for risks associated with ***delivery***.

This includes:

- Manage the delivery of software solution
- Matches skills/resources, sets technology boundaries & engineering practices for system evolution & integrity
- Provides guidance, looks for risk, establishes technical boundaries to empower the team
- Supports the process
- Remove impediments

The TDM is interacting with the Business to assure the right work being done at the right time to achieve a Business objective: prioritization, decomposition, impediments and cross-impacts being addressed, is work ready to be pulled by the team. The TDM is empowered to make decisions about work prioritization based on technology requirements that affect flow.

The TDM is the only person to go to for status of the project. While the Product Owner can answer questions about whether we are delivering value to the Business, the TDM gives status on everything else such as:

- What is the timeline for delivering features
- Will we make the timeline
- What is getting in the way

These responsibilities must be done... Who does them?

Who should play the Application Development Manager and Technology Delivery Manager roles? This is not quite the correct question because a "role" is just shorthand for a set of responsibilities. The most important thing is to understand the responsibilities described in the ADM and TDM roles and then to be sure that these responsibilities are being attended to. How you divide the responsibilities depends on the context of your organization. In larger organizations, you may have dedicated people to play the ADM and TDM roles while in smaller organizations, the responsibilities might spread across several positions.

Who does the work? We have seen the following:

- TDM responsibilities being done by senior project managers
- ADM responsibilities being done by architects who also have management responsibility, development managers, and even leads.

The larger you get, the more likely it is that you will need dedicated ADM and TDM roles. The systems are more complex and so require dedicated attention to the issues.

Whether an organization is four people or four thousand, organizations who are trying to be serious players are driving to increase the maturity of processes and practices, quality, metrics. It means they must have standards and then enforce those standards. The sets of responsibilities contained in the ADM and TDM responsibilities, together, assure consistency and growth.

Assuring that both sets of responsibilities are being attended to has these benefits:

- Teams are properly focused and prepared to deliver value
- Status meetings are shorter and more effective, with a clear delineation between value (reported by the Product Owner) and project (reported by the TDM)
- Forecasting and predictability is better
- Quality improves because of fewer bugs, better testing, more complete Done criteria

It is possible that an organization might achieve success without these roles in the short term. However, any organization, regardless of size, must start paying attention to these responsibilities in order to achieve on-going success in the long term. It is easy to succeed at release 0.1. But the bulk of your work and the bulk of the value you deliver will be in the future releases. The ADM and TDM roles assure ongoing quality and integrity and flow for these future releases. Viable, on-going product development requires investment in these roles.

Viable, on-going product development requires investment in these roles.

Business Product Owner

The Business Product Owner role is responsible for the delivery of a successful product.

The Business Product Owner holds primary responsibility for representing two key stakeholders: the Business and the customer. She sets the vision for what to build. To accomplish this, the Business Product Owner collaborates with all appropriate stakeholders to establish and communicate the product vision, product road map, and the prioritized backlog. This requires close involvement with the business stakeholders to understand opportunities and manage expectations.

The Business Product Owner either is or represents the primary business sponsor for the project. This requires an ability to represent the product to the various Business stakeholders and sponsors, securing funding, and defending it from inappropriate interference.

The Business Product Owner is unafraid to *set direction* for the product and humble enough *not to tell how* to do it.

The Business Product Owner works closely with the ADM, TDM, Scrum Master, and the team to ensure that the product evolves at a challenging but sustainable pace and that progress is visible. They also work together to help the team resolve impediments, taking ownership for impediments that lie outside the team's span of influence. The Business Product Owner must have a good sense of the team's capabilities.

In the day-to-day work, business analysts often represent the Business Product Owner. It is not uncommon for the Business Product Owner to have some management responsibilities for business analysts on the team. However, it is critical that they speak with one single voice.

Accountability	Which involves...
During Planning	Identify features to be delivered by developing an understanding of the product stakeholders and their needs as they relate to the product.
	Develop, communicate, and evolve the product vision, bringing context to the work.
	Identify and communicate business milestones, dependencies, and risks.
	Prioritize and manage the product road map and backlog based on business value.
	Develop and communicate release goals and iteration goals(s).
	Identify prioritized candidate stories for the next iteration from the product backlog.
	Review with team to identify required additional stories.
	Along with the team, commit to the iteration.
Daily Work	Attend Daily Stand-ups when needed.
	Provide clarity on the stories and associated test and acceptance criteria.
	Ensure that preparation for future iterations is planned by adding "vision creation" stories to active iteration.
	Adjust commitment total (top-line) with the team to fit within the iteration.
	Monitor release and iteration burn-up and make adjustments as needed.
	Monitor product business value progress and feature burn-up and make adjustments as needed.
	Manage the product rollout strategy and implementation approach.
	Manage the feedback loop ensuring the business and development team stay synchronized.
	Help resolve impediments that lie outside the team's span of influence.
Iteration Closing	Participate in iteration retrospections.
	Review product / product increments with other appropriate stakeholders.
	Review actionable improvement goals.

Scrum Master

The Scrum Master role is responsible for

- Ensuring the process is understood and followed
- Facilitating the teams efficiency and health
- Ensuring visibility on progress
- Providing transparency of process and work in progress

The Scrum Master champions the needs of the team to the organization and for championing the needs of the organization to the team. The Scrum Master is the team's facilitator, shepherding the team as a servant-leader by creating a positive and safe environment, improving team cohesion, being creative and focused, asking difficult questions aimed at challenging the team, and bringing issues and impediments to the surface.

The Scrum Master helps the Daily Stand-up stay on track and is the primary source for proactively radiating project information. As a practical reality, the Scrum Master is usually the role responsible for maintaining the team's project board, ensuring it is up to date before the Daily Stand-up. In addition, if the team uses a project management portal, the Scrum Master usually is responsible for maintaining it, implementing security and accounts, creating reports, etc.

The Scrum Master helps remove impediments and keeps the product and the process moving forward.

In some organizations, project managers play the role of Scrum Master; however, they must be careful to focus on improving the team's process over against just schedule and budget.

Accountability	Which involves....
During Planning	With all appropriate stakeholders, prepare for the upcoming iteration planning session. Address: • Product vision • Product backlog • Risks / Dependencies • Release plan • Artifacts and deliverables for the project • Iteration length Support the team's ability to commit by communicating team capacity. Negotiate commitment between the Business Product Owner and the team. Help team create information visibility boards. Establish the progress charts: Product burn-up, Iteration Burn-up and Burn-down. Agree on Daily Stand-up location and times. Set up work areas. Help the team develop communication rules. Update spreadsheets and database information. Show number of points that have been committed to for the iteration.

Scrum Master

Accountability	Which involves....
Daily Work	Reflect on how the iteration is going, looking for tasks taking too long.
	Facilitate Daily Stand-ups. Ensure that people clearly address the team and stay focused.
	Radiate information to show progress clearly and predictably.
	Update visibility tools with the latest task and story progress.
	Post updated charts prior to the Daily Stand-up, current as of yesterday.
	Identify and display all impediments and issues that affect team velocity and iteration successes.
	Clearly show activity and progress for impediments resolution.
	Engage team members to identify impediments, build rapport, and improve team cohesion.
	Shield the team from external interfaces and distractions.
	Facilitate team meetings and process improvement workshops.
	Challenge the team and to contribute fully while remaining aware of team cohesion and sustainability.
	Uphold the Agile process and challenge the team to identify and implement process improvements.
	Socialize Agile within the organization.
Iteration Closing	Help demonstrate product increment to the Business Product Owner and other appropriate stakeholders.
	Facilitate iteration retrospection. Identify actionable improvement goals for the next iteration. Place improvement stories into product backlog.
	Share the findings from the iteration retrospection, including the process improvement stories, with other Scrum Masters in the organization so that the knowledge can be applied more widely and refined as more is learned.

Business Analyst (Business SME)

The business analyst role represents the Business to the team. This includes describing business knowledge and business rules, answering questions about customers, performing analysis that will become stories, and writing acceptance tests.

The business analyst is a member of the team, swarming on work items together with other members of the team if they have skills that will help the story to be completed. Business analysts are commonly involved in acceptance testing.

The whole team drives the product from a tactical perspective. It is autonomous and self-organizing and is accountable to the Business Product Owner for committing to and delivering product increments within the duration of the iteration.

The whole team works in parallel with all appropriate stakeholders to estimate the size of backlog items, make design and implementation decisions, and transform stories into high quality software assets.

Working with the Scrum Master, the team tracks its own progress, raises awareness for impediments, and devises options to maintain forward progress. The team is responsible for establishing processes and standards, continuously improving them, and then following those processes and standards.

Accountability	Which involves...
During Planning	Participate in sizing, prioritizing, and decomposing stories into implementation stories and tasks.
	At a high level, identify enabling and systematic evolution issues that must be addressed.
	Identify stories which might need to be addressed in the upcoming iteration.
	Identify all skills necessary to achieve the iteration commitment.
	Collaborate with Business Product Owner and commit to delivering an increment of the product software.

Accountability	Which involves...
Daily Work	Focus efforts on delivery of the highest priority stories and tasks while identifying dependencies.
	Build rapport with team members to increase team collaboration.
	Maintain a positive and safe work environment, and increase trust.
	Contribute to development and testing by swarming with other team members on stories and tasks.
	Help resolve impediments that are within the team's span of influence.
	Contribute to the Daily Stand-up.
	Clearly mark effort on remaining tasks and who the owner is.
	As necessary, rewrite tasks to level workload, handle dependencies and impediments, accelerate results.
	As necessary, rewrite stories to level work, handle dependencies, and impediments, accelerate results.
	As necessary, resize stories based on any new information or what has been learned.
	Discuss issues raised during the Daily Stand-up about which the analyst can help.
	Help team contribute while remaining aware of long-term team cohesion and sustainability.
	Ensure the test strategy is implemented effectively and in a timely manner.
Iteration Closing	Demonstrate product / product increment to Business Product Owner and other appropriate stakeholders.
	Participate in the iteration retrospection. Identify and commit to actionable improvement goals for the next iteration based on what worked well and what can be improved.

Developer

The developer role is responsible for working with other members of the team to build potentially releasable features in every iteration. The developer collaborates with business analysts and testers on stories and tasks, ensuring that the results meet all functional and acceptance tests. The developer also helps the team resolve impediments, getting help as needed.

The whole team drives the product from a tactical perspective. It is autonomous and self-organizing and is accountable to the Business Product Owner for committing to and delivering product increments within the duration of the iteration.

The whole team works in parallel with all appropriate stakeholders to estimate the size of backlog items, make design and implementation decisions, and transform stories into high quality software assets.

Working with the Scrum Master, the team tracks its own progress, raises awareness for impediments, and devises options to maintain forward progress. The team is responsible for establishing processes and standards, continuously improving them, and then following those processes and standards.

Accountability	Which involves...
During Planning	Size, prioritize, and decompose features into stories and tasks.
	At a high level, identify enabling and systematic evolution issues that must be addressed.
	Identify process and environment issues which might need to be addressed in the upcoming iteration.
	Identify all skills necessary to achieve the iteration commitment.
	Collaborate with the Business Product Owner and commit to delivering an increment of the product software.

Business Analyst (Business SME)

Accountability	Which involves...
Daily Work	Focus efforts on the delivery of the highest priority stories and tasks while identifying dependencies.
	Collaborate and build rapport with other team members to increase team cohesion.
	Maintain a positive and safe work environment, and increase trust.
	Contribute to development and testing by swarming with other team members on stories and tasks.
	Develop tests and code that conform to the code quality standards agreed to by the team.
	Help resolve impediments to progress that are within the team's span of influence.
	Contribute to the Daily Stand-up.
	Clearly mark effort, ownership for remaining.
	As necessary, rewrite tasks to level workload, handle dependencies and impediments, accelerate results.
	As necessary, rewrite stories to level work, handle dependencies, and impediments, accelerate results.
	As necessary, resize stories based on any new information or what has been learned.
	Discuss issues raised during the Daily Stand-up about which they can help.
	Help the team contribute while remaining aware of long-term team cohesion and sustainability.
	Ensure the test strategy is implemented effectively and in a timely manner for SAT, CAT.
Iteration Closing	Demonstrate the product / product increment to the Business Product Owner and other stakeholders.
	Participate in the iteration retrospection. Identify and commit to actionable improvement goals for the next iteration based on what worked well and what can be improved.

QA / Tester

A key Lean-Agile principle is to aim for perfection, to improve constantly. It is everyone's job. This informs role of the testing: it is done early in the process and is the responsibility of the team. Every feature and every story must have an acceptance test. The outcome of testing is to deliver fairly well-perfected code where defects have no serious consequences.

Quality assurance (QA) does help discover bugs. But more importantly, the QA/Tester role is a full partner in the team, to help discover the causes of errors and eliminating them. This involves root-cause analysis, looking at processes, infrastructure, and the understanding of customer requirements.

A person in the QA/Tester role must understand the context for the project and help others to make informed decisions based on this context. A key responsibility is to help the business analyst create acceptance tests up front.

The QA/Tester helps set up a testing environment that is consistent with the team's quality objectives. The tester then works with developers and business analysts to help them create and conduct tests with complete coverage.

Accountability	Which involves....
During Planning	Size, prioritize, and decompose features into stories and tasks.
	At a high level, identify enabling and systematic evolution issues that must be addressed.
	Identify process and environment issues which might need to be addressed in the upcoming iteration.
	Identify all skills necessary to achieve the iteration commitment.
	Collaborate with the Business Product Owner and commit to delivering an increment of the product software.

Accountability	Which involves....
Daily Work	Focus efforts on delivering highest priority stories and tasks while identifying dependencies and risks.
	Collaborate and build rapport with other team members to increase team cohesion.
	Maintain a positive and safe work environment, and increase trust.
	Contribute to development and testing by swarming on stories and tasks.
	Develop tests and code that conform to agreed upon code quality standards and practices.
	Work with team to create the build environment, with special emphasis on the testing environment.
	Help resolve impediments to progress that are within the team's span of influence.
	Contribute to the Daily Stand-up.
	Clearly mark effort on remaining tasks and who the owner is.
	As necessary, rewrite stories to level work, handle dependencies, and accelerate results.
	As necessary, resize stories based on any new information or what has been learned.
	Discuss issues raised during the Daily Stand-up about which they can help.
	Help team contribute while remaining aware of long-term team cohesion and sustainability.
	Ensure the test strategy is implemented effectively and in a timely manner: SAT, CAT, end-to-end.
Iteration Closing	Demonstrate product / product increment to Business Product Owner and other appropriate stakeholders.
	Participate in the iteration retrospection. Identify and commit to actionable improvement goals for the next iteration based on what worked well and what can be improved.

Release Manager

The Release Manager role is different from traditional approaches. Because Lean-Agile focuses on smaller, and perhaps fewer, features within a release, the need for complex configuration and release management is less involved.

In some organizations, perhaps especially in organizations developing software for external customers, release management is a separate and specialized position. In other organizations, the role is handled by the Business Product Owner or a project manager. However it is staffed, the important point is that the responsibilities of this role are critical and must be handled by someone.

The release manager is responsible for helping the Business Product Owner understand the required features that must go into the "Minimal Business Increment" set (the minimum set of product that must be realized in order for the customer to perceive value). This involves understanding and advising about dependencies by or on other products being developed across the enterprise. It also involves representing the requirements of packaging, technical support, and physical distribution of the product.

On occasion, the Business Product Owner and the Team may find themselves in an adversarial relationship. There may be problems in communication or lack of understanding of business needs or difficulty explaining true technical requirements or sorting out a team's desire to do something "cool." In such a situation, the release manager role may be called upon to help mediate.

The Release Manager helps to resolve impediments, particularly which pertain to support, marketing, packaging, and related release activities.

Accountability	Which involves...
During Planning	Size, prioritize, and decompose features into stories and tasks.
	At a high level, identify enabling and systematic evolution issues.
	Identify process and environment issues which might need to be addressed in upcoming iterations.
	Collaborate with Business Product Owner and commit to delivering an increment of the product software.

QA / Tester

Accountability	Which involves...
Daily Work	Focus efforts on delivery of the highest priority stories and tasks while identifying dependencies.
	Build rapport with other team members to increase team cohesion
	Maintain a positive and safe work environment, and increase trust.
	Contribute to development and testing by swarming with other team members on stories and tasks.
	Help resolve impediments to progress that are within the team's span of influence.
	Contribute to the Daily Stand-up.
	Clearly mark effort on remaining tasks and who the owner is.
	As necessary, rewrite stories to level work, handle dependencies, and impediments, accelerate results.
	As necessary, resize stories based on any new information or what has been learned.
	Discuss issues raised during the Daily Stand-up about which they can help.
	Help team contribute while remaining aware of long-term team cohesion and sustainability.
	Ensure that the release strategy is implemented effectively and in a timely manner.
Iteration Closing	Demonstrate product / product increment to Business Product Owner and other appropriate stakeholders.
	Participate in the iteration retrospection. Identify and commit to actionable improvement goals for the next iteration based on what worked well and what can be improved.

Database Administrator

The main goal of the database administrator role is to support the creation of database projects as well as the deployment to production of database project changes. This is in addition to the database administrator's traditional role of performing day-to-day administration and maintenance of database servers.

The database administrator is not expected to participate in the Daily Stand-up or retrospections unless a member of the team.

Accountability	Which involves...
During Planning	Advise the database administrator, identify enabling and systematic evolution issues that must be addressed for data environments. Help to identify all skills necessary to achieve the iteration commitment.
Daily Work	Focus efforts on the delivery of the highest priority stories and tasks that pertain to the database and data environment while identifying dependencies to watch out for.
Iteration Closing	Participate in demonstration of product / product increment to Business Product Owner and other appropriate stakeholders.

Database Developer

The database developer's main goal is to implement all database development tasks within the planned time frame. The database developer is also responsible for cost estimation, supervision of feature implementation, and providing database expertise to developers on the team.

Database resources are often quite limited. It is common in Lean-Agile projects for a lead developer to assume the role of a "surrogate database developer/analyst." This role acts as the primary intermediary between the team and the database developer, collecting requirements, managing the relationship, and reporting progress. This approach creates higher quality and higher bandwidth communication with the limited resources.

The database developer or surrogate participates in the iterative database development life cycle, along with the database administrator and the team members.

The database developer is not required to participate in the Daily Stand-up, but may participate in retrospections. The database developer may play an advisory role in story planning.

Accountability	Which involves...
During Planning	Advise the database administrator and Technical Champion; identify enabling and systematic evolution issues to address for data environments.
	Help to identify all skills necessary to achieve the iteration commitment.
Daily Work	Focus efforts on the delivery of the highest priority stories and tasks that pertain to the database and data environment while identifying dependencies to watch out for.
	Ensure all database development tasks are implemented within planned time frame.
	Provide cost estimates.
	Provide expertise to developers and analysts.
Iteration Closing	Demonstrate product / product increment to appropriate stakeholders.
	May participate in retrospections.

A Team of Peers: Roles in Lean-Agile

This team of peers is founded on the following core principles:

Principle	Description
Team of peers	A team of peers with a common focus, shared responsibility and open communications. Each role is accountable for a specific share of the quality of the overall solution.
Roles and skills	Team membership is based on roles and skills rather than positions. People have a primary role and periodically occupy other roles on a team depending on their skills and the needs of the release.
Delivering value	A focus on delivering value as prioritized by the Business. The team works from a common vision for the product and for releases, as laid out by the Business Product Owner. The goal for each release is to deliver the minimal set of releasable product that will help the Business. Teams commonly swarm on stories to focus on one or just a few stories at a time.
Frequent feedback	Get feedback early and often. The team's activities are set up to get feedback from all stakeholders to help them learn what their requirements are. The idea is to work on what is known rather than making customers speculate about what they do not yet know. Alternatives and changes are much easier earlier than later.
High-bandwidth communication	High-bandwidth communication is preferred to encourage fidelity and speed in decision-making and to minimize documentation that is truly non-value-added. Teams use tools to promote this communication.
Deliver product	The team is focused on delivering product. The primary focus of the team is a working product that delivers the features required by the Business Product Owner. That is what they commit to and they are expected to deliver.

Principle	Description
Team owns their process	The team is responsible for their process. Teams employ their local knowledge about what is required to do their work, complementing enterprise standards that are required. The team agrees to use and improve its own processes. If something is not working, stop, improve, and then go forward.

Swarming and Teamlets

The team's focus is on building and delivering product. People are expected to bring their skills to bear, regardless of role, to work on whatever is currently active. For example, a Business Analyst might volunteer to help with acceptance testing of a piece of functionality.

This idea of people joining to work together based on their skills rather than their roles is called "swarming" and is an essential team skill. When people agree to swarm on a work item, they form a "teamlet." The teamlet may assign a "Story Captain" to help the teamlet stay on track.

Teamlets are generally formed at the Daily Stand-up, when the team reviews progress that has been made, and decide what needs to be worked on next. They are fluid, forming and disbanding based on the requirements of the work items at hand.

Swarming Rules	Description
Focus on only one story at a time	Within an iteration, teams should have only a few stories open at a time because the focus is on burning down stories. Do not dissipate energy by focusing on too much at once.
The swarm is the priority	While individuals may work on other tasks, their priority should be the stories they swarm on.
If you have the skill, join the teamlet	If you have the skills to contribute to burning down a story, and you have capacity, you are expected to join in the teamlet, even if it is not exactly in your job description to do so.

Part III: From Concept to Consumption

In Scrum, work proceeds from the initial concept through development, and on to delivery when the customer can consume the product.

This section has three parts

- The activities of discovery and delivery involved
- The elements used in these activities: Vision, Features, Stories, and Tasks
- Estimation

Overview: Business Discovery and Business Delivery

In Scrum, planning and re-planning is done continuously along the entire value stream from vision to support. Lean-Agile principles help Scrum to focus this by driving all effort to delivery Business value.

The goal is to plan as much as the Business customer is clear about and to avoid guessing while incorporating lessons learned and customer feedback. This avoids building features that are not (yet) needed or building features with the wrong behavior. It recognizes that users often do not know what they need until they have something concrete to work with. It allows plans to adapt to changing market conditions and business priorities. And, it allows teams to deliver sooner something that the customer finds valuable rather than having to wait.

> *In preparing for battle, I have always found that plans are useless, but planning is indispensable.*
> *– Dwight D. Eisenhower*

Phases of Business Discovery and Business Delivery

Lean-Agile describes the value stream of discovery and delivery in eight phases with four gates. These are illustrated in the arrow figure. The outcomes for the stages are described in the following table.

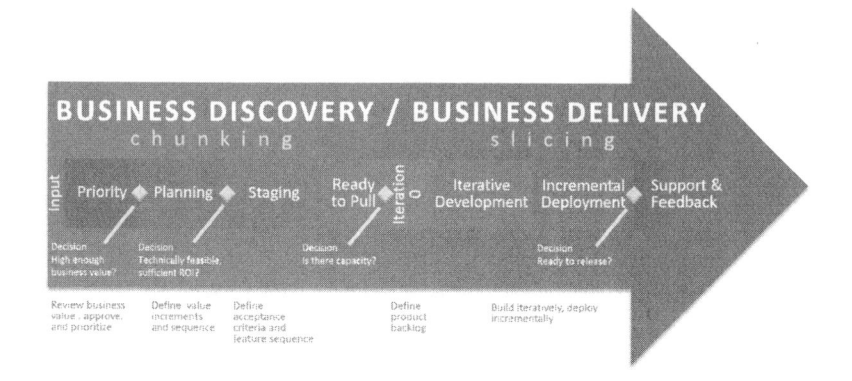

BUSINESS DISCOVERY / BUSINESS DELIVERY
chunking — slicing

Input → Priority ◆ Planning ◆ Staging → Ready to Pull ◆ Iteration 0 → Iterative Development ◆ Incremental Deployment ◆ Support & Feedback

Decision: High enough business value?
Decision: Technically feasible, sufficient ROI?
Decision: Is there capacity?
Decision: Ready to release?

Review business value, approve, and prioritize
Define value increments and sequence
Define acceptance criteria and feature sequence
Define product backlog
Build iteratively, deploy incrementally

Stages and Outcomes of Discovery and Delivery

Stage	Responsible	Outcomes
Priority	Business Sponsor Business Owner Stakeholders Business PM	Business Value established Approved to Continue Prioritized (relative to other approved items) Product Owners identified Book of Work updated Product Vision
Planning	Business Owner POs Business PM Technical Owner ADMs / Leads	MBIs defined Approved to continue Prioritized and sequenced based on ROI (relative to other approved items) PO assigned Business SMEs identified Book of Work updated

Overview

Stage	Responsible	Outcomes
Staging	Business Owner Product Owners Business SMEs Business PM Technical Owner ADM / Leads	MBI(s) refined, sized, sequenced based on ROI Features defined and sequenced Acceptance criteria defined Business Backlog established
Ready to Pull	Business Owner Product Owners Business PM TDM	Business Owner available Product Owners available Business backlog defined TDM assigned / signed off Required resources available Team(s) identified with available capacity
Iteration 0	Business Owner Product Owners Business PM Customer / User Representatives Technical Owner ADM / Tech Leads Representatives for all skills required Scrum Master(s)	Program backlog established for all work: Business and technology Feature sequence refined including technical dependencies Initial top-line in story points High level architectural specified Done: Criteria, tests, and documentation Logistics: Team(s), visual controls, meetings, coordination
Iterative Development	Business Owner Product Owners Customer / User Representatives TDM / PMs ADM / Tech Leads Representatives for all skills required All team members Scrum Master(s)	Iteration Backlog established of stories committed by the teams for next 2 weeks Right-sized stories defined, sequenced, sized in story points Tasks defined in hours Initial Iteration burn-up and burn-down reports

Progressive Unfolding

Lean-Agile thinking, analysis involves a progressive unfolding of detail from product vision to features to stories and specific tasks. Each level of detail describes with more certainty what the software product development team needs to do, when they need to do it, and how they will know when they have succeeded in order to deliver value to the Business. At each level of detail, it becomes easier to estimate the work involved.

Planning and analysis are activities the team uses to foster *communication* between team members, stakeholders, and customers.

Planning and analysis cannot be separated cleanly in Scrum. Each drives the other and each provides feedback. Teams constantly re-plan and add detail based on what has been learned. Give the team just enough so that they can start working on delivering value. *Never speculate* about requirements.

Lean-Agile thinking , analysis involves a progressive unfolding of detail from product vision to features to stories and specific tasks. Each level of detail describes with more certainty what the software product development team needs to do, when they need to do it, and how they will know when they have succeeded in order to deliver value to the Business. At each level of detail, it becomes easier to estimate the work involved.

Planning is the responsibility of the Business Product Owner who represents the voice of the customer and the voice of the Business to prioritize the work to be done. Planning is done all throughout product development:

- Product visioning
- Release planning
- Iteration planning
- Daily planning with the Daily Stand-up

Analysis involves defining features, stories, and tasks, the essential elements Scrum teams use to do work.

Activities of Business Discovery: Priority, Planning, and Staging

Priority, Planning, and Staging are the stages of Business Discovery. The goal is to understand what is required by the Business and to prioritize, sequence, and prepare work so it is ready to pull by the team.

Business Discovery is beyond the scope of this pocket guide. This section offers some important concepts that occur in Business Discovery.

Checklists

Checklists for Business Discovery begin on page 113.

Prerequisites

Before planning can commence, the leaders for the project must be selected. This would include identifying the Business Product Owner, Scrum Master, and lead architect. Typically, the team (or most of it) will also have been identified. The product vision should be understood.

Activities

The activities of planning and staging include the following:

- Overall estimation and prioritization
- Identify impediments to remove and analysis to conduct to allow work to progress
- Identify lessons from previous releases or iterations that the team wants to incorporate into this release
- Review and revise the team's standards and practices

About Estimation and Prioritization

During planning, estimation and prioritization involves three steps.

Step	Do this...
Assigning business value	Assigning business value is done by the Business Product Owner. It is done independently from estimating how much effort will be involved in building the feature or story. Business value is the value that will be received after the feature or story is built. It should use the same scheme used for Story Points.
Estimation	Estimation is a measure of the size, in points, of the feature or story to be built. The Team Estimation Game, described above, is an effective approach to estimation.
Prioritization	After the business value has been established and the size of the feature or story has been established, prioritization can take place by looking at return on investment.

Estimating Large Features

Estimation of large features is often done by comparing one feature to another. If one is 300 points, how does it compare with another feature?

Clarity on Estimation

The estimate for story size needs to be for completing the story. That is, for it to be taken to "done, done, done": not merely coded, but accepted by the customer. Agile methods require that stories are completed each iteration. Not getting through test and validation means the story is not complete.

Issues and Considerations

Topic	Discussion
Co-location	People communicate much more efficiently when they are together. Try to co-locate during Iteration 0 whenever possible.

About Planning and Staging

Product planning and staging is one of the most important – and complex – activities in Lean-Agile. The Business Product Owner together with key business stakeholders and technical subject matter experts (SMEs) must determine what is required in order to realize the product vision.

Product planning involves satisfying a number of objectives:

- Achieve the business case that justifies the effort
- Satisfy the most valuable sets of users as soon as possible
- Deliver the functionality that users require in order to do the work they require
- Ensure technical integrity with the overall code base
- Minimize risk due to changes in to the system
- Maximize sustainability over the life of the product

In Lean-Agile, manifesting the product vision is done incrementally. Every increment to the product is called a "release." A release includes both the technical functionality and the processes, training, and support materials required to make the functionality useful for the target set of users. Releases are developed iteratively.

The Business Product Owner must orchestrate the various activities of planning so that the business case that justifies and governs the product work is achieved as *soon* and as *profitably* and as *responsibly* as possible. This plan is continuously evaluated to ensure that effort is progressing well, based on what has been learned along the way.

The release plan makes the decision to release easy: if the technical functionality has been accepted by the customer (it is "done") and has integrated successfully with the production system *and* the target users are ready to use the new functionality *and* the team has held retrospection on their process to incorporate lessons for the next cycle of work, then the release is complete.

Checklists

A checklist for planning is on page 113.

A checklist for staging is on page 114.

Prerequisites

Before the product plan can be created, the Business must specify initiatives for the value stream. The Business Product Owner assumes responsibility for planning on behalf of the Business.

Activities

In product planning, the planning team does the following:

- Understand the goals of the initiatives and how these are expressed in features
- Develop general release plan, including elevations and usage
- Identify impediments to remove
- Identify lessons from previous releases or other projects that the team wants to incorporate.
- Ensure proper feedback: product demonstration to stakeholders and review of standards and practices by the team

The Thinking Activities of Planning and Staging

Product planning and staging is complex. It involves four types of thinking activities that must work in harmony for the purpose of achieving the ultimate business case: Analysis, selecting, adjusting, and ordering. The Business Product Owner coordinates these thinking activities, deciding when to shift focus from one to another in order to minimize effort, mitigate risk, and handle uncertainty so that just the right amount of detail is being achieved.

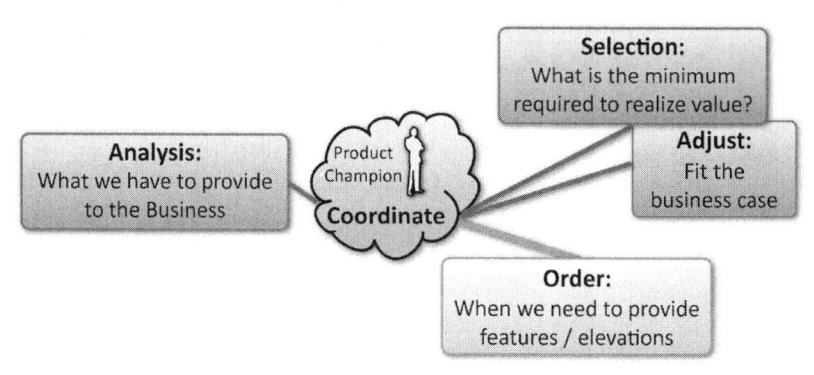

Analysis: What we have to provide to the Business

Product Champion — Coordinate

Selection: What is the minimum required to realize value?

Adjust: Fit the business case

Order: When we need to provide features / elevations

Five tasks of product planning and staging

Activities of Business Discovery

Activity	Brief Discussion
Analysis	Analyze the product vision, breaking it down into finer and finer detail: Capabilities, features, stories, and tasks so that: • The team knows what to do and how to validate they have done it • The Business knows the changes to process and people required to make functionality usable by target users At any stage, decompose just enough to be able to minimize risks and to plan what is going to be done in the near term.
Selection	What is the most *responsible way* to achieve the *minimum capability* required to satisfy the business case? Such that the target user set can begin using it profitably? From the list of minimum feature sets, what is the minimum set of elevations required to ensure technical integrity. Change management is involved early in this process, focused on both technical and business/use.
Adjust	Do the minimum sets fit the business case that is justifying the work? Are we meeting the requirements of the most important target user set? Does the work plan fit within the business case? Does it consume too much of the business case so that other work will not get done and defeat the overall goals? Should the minimum sets be adjusted in order align more perfectly with the business case?
Order	Sequence the minimum feature sets and associated elevations so that work is completed and payback is achieved in a timely way, according to what the target user set can consume profitably.
Coordinate	Manage these analysis and planning activities from a global project and business perspective: • *Minimize the effort*: both the effort of planning itself and the resulting work that the team will be required to do during the next planning horizon • *Mitigate risk*: giving attention to risks to project and system that must be addressed early • *Manage uncertainty*: plan to maximize learning and feedback, help the team to guess well.

Analysis: What we have to provide the Business

Analysis is the analytical process of unfolding the details required to realize the vision.

Description	Example: Web Self-Service
Business Capability. Business capabilities include both the technical capabilities of the product and the people skills and processes required so that customers can work with the product.	Customer Self-Service over the web.
Feature. Describes in business terminology the essential elements required to realize a capability. Features are essential units for releases: releases are loaded up with features until it meets the whole team's capacity.	Account management over the web
Story. Describes fundamental increments of business value together with the validation criteria to know when that increment is done. Stories are tied to features. In release planning, the release planning team uses stories to help understand what is involved in realizing the feature and the relative size of the feature: sizing stories is easier than sizing features since they are smaller. During Iteration 0, the whole team will add to the list of stories based on what they have learned. And they will decompose stories until they are the right size to be completed within an iteration, usually in 1-3 days.	Create an account Read an account Update an account Delete an account (requires support-desk assistance)

> *Prioritization does not mean merely diving stories into "must have", "important to have", "need to have", ... This has too few buckets. It doesn't pressure the Business Product Owner into rank ordering the features. Prioritizing features means most important, then next most important, and then the next most important, ...*

Selecting: The minimum required to realize value

Product planning begins with an understanding of what the business wants to deliver to customers. Realizing this product vision is done incrementally because everything cannot be delivered at once. Each increment contributes to the vision, contributes value to the Business. Each increment to the product is called a "release."

Delivering product to the Business incrementally means that the Business is always realizing value as quickly as it can absorb it. Smaller units of work mean the whole team can deliver more product that is tuned to the needs of the business. It allows the Business to drive the software product development according to its needs rather than according to the needs of the system.

A release puts the product to use in order to achieve business value. It includes both *technical functionality and integrity* and *business use:* both the technical work to introduce functionality into the product or systems *and* the materials and processes required to make the functionality ready to use by a target set of users: training, documentation, support knowledge base articles, certification, pricing, packaging, promotion, distribution plans, and so forth.

This involves defining the set of **Minimal Business Increments** (MBI) required to realize business capabilities.

- *Minimal* means the targeted set of users will receive the necessary and sufficient capability to realize value from using it.
- *Business Increment* means the combination of functionality and associated business use materials and processes required for a set of users to realize business value.

Think of it like this:

- Delivering less will not help them do the work they need to do.
- Delivering them more might delight them but if there is not a compelling marketing reason to do so, focus on other features that have a higher priority.
- Delivering too much or more than the customer can absorb and use is wasted effort.
- Delivering to the wrong set of users means under-serving higher priority users (as defined by the Business).

The Business defines these minimum sets based on their objectives, such as expanding market segmentation, regulatory compliance, or product line differentiation.

There are many strategies for selecting MBIs. Here is a good one:

Step	Description
Define actors	Define the set of "actors" – people, roles, other systems – who will use the product. Are any of these of higher priority to the Business?
Define scenarios	Look at how a particular set of actors interact with stories in order to realize a particular capability. There may be many scenarios.
Package into capability sets	Based on business objectives, package the scenarios into sets of capabilities that will be valuable for a target set of users.
Define MBIs	For each capability set, identify the features required to realize the capability. Package the features into Minimal Business Increments which can be released to be useful for the customer.

Adjusting: Releases and the Business Case

Releases must support the business case that justifies and guides the effort. The best release is the one that best satisfies the business case. Here are two situations that might require adjusting the MBI:

- **An MBI consumes too much work.** An MBI that consumes too much of the budget is too big if it means that other MBIs which are also required cannot be done. Focus on a smaller target user set.

- **The business case is achieved.** If the Business judges that a release adequately satisfies the business case, they may choose to stop work. The remaining MBIs may no longer be necessary. Redeploy the team to other, higher-priority work.

Adjusting the MBI set is done throughout the life of the project as more is learned.

Selecting: Elevations

Perfecting the technical aspects of the system is crucial. Each iteration involves assuring the technical product integrates with the rest of the production system correctly and without causing harm.

The "elevation" is a planning tool that helps the team plan work iterations that minimize risk, increase learning, and coordinate effectively with other teams. Given all of the work required to realize an MBI, the team defines an objective for a particular iteration or set of iterations. The elevation is what is required to achieve that objective and ensure that the resulting code integrates well and maintains technical integrity.

Example. A particular elevation may focus on one or a few architectural and risk issues. As they plan the iteration, the team focuses on selecting all the stories that associate with those issues.

Elevations help make visible to the team and to the Business what is going on in an iteration.

Using minimally-sized elevations minimizes the risk to the system by reducing the amount that is changing at any time.

Elevations are not releases. An elevation focuses on the technical integrity part of a release. It can make the functionality available to the wider development organization and even to a select set of customers. It is not a "release" because it does not involve making the functionality usable for all target customers.

> *What is the best approach for elevation?*
>
> *The team must decide its own approach to elevation. Elevation may use Continuous Integration (CI) where code is integrated continuously or nearly continuously; however, this is not always possible. The process of integration may be too costly, may involve multiple teams, or production environment may not allow it. In such cases, teams may have to integrate in batch or even use a separate elevation iteration. But, elevation iterations may indicate you are not detecting errors early enough - you want to correct that.*

Ordering: When we need to provide it

Determine when to release various sets of MBIs to customers. The Release plan lays out the priority and expected time intervals (e.g. monthly, quarterly, as finished) for delivering MBIs.

There are many strategies for doing this; here is one good approach:

Step	Description
Prioritize MBIs for value	**The Business prioritizes the MBI, high to low.** The goal is relative ranking of MBIs: is *this MBI* relatively more important than *that* MBI? Factors to consider include:
	Mitigate risk and uncertainty. Favor MBIs that deliver what customers are certain about. Favor MBIs that exercise risky parts of the solution. Favor MBIs on which other MBIs depend.
	Maximize ROI. Favor MBIs that equip the Business to do high-value work. Favor MBIs that customers strongly identify as required or satisfiers.
	Manage effort by team capacity. Favor MBIs that meet the team's capacity.
	Note: The Team Estimation Game is helpful here.
Fill the release	The team has the capacity to handle a certain number of feature points. Fill the release with features up to the capacity of the team.
Adjust the plan	The release plan is adjusted as more is learned about the customer, product, and team capacity.

Avoid failure to launch!

The Business focuses on the release: when the feature will be usable. They are driven by business objectives and the business case.

IT/Development focuses on elevations: assuring technical integrity. They are driven by system issues.

The Change Management role is responsible for addressing both dimensions. This focus is how you avoid failure to launch.

Issues and Considerations

Release Planning

Topic	Discussion
Avoid failure to launch	The Business focuses on the release: When the feature will be usable. They are driven by business objectives and the business case. IT/Development focuses on elevations: Assuring technical integrity. They are driven by system issues. The change management role is responsible for addressing both dimensions. This focus is how you avoid failure to launch.
Velocity is unknown at the start	The capacity of the team and the ability to estimate are usually not well understood at the beginning. It may be +/- 100% the first time through. Give yourself grace.

Elevation

Topic	Discussion
Unclear feature complexity	If there is very little known about the cost of a feature, then you will have to drive down with spikes or such.
How to elevate?	The team must decide its approach to elevation. Elevation may use Continuous Integration (CI) where code is integrated continuously or nearly continuously; however, this is not always possible. The process of integration may be too costly, may involve multiple teams, or production environment may not allow it. In such cases, teams may have to integrate in batch or even use a separate elevation iteration.
Disable features	Features that are integrated into the production code must never be accessible to users or systems until the features are released.
Alert the Business	Still, bad things can happen. It is possible that an integration may cause harm inadvertently. It is crucial that the Business and the support desk know about integrations that are going in.

Iterations

Agile uses an iterative approach to product development: building product features in 2-4 week increments (iterations). Iterations begin after the Iteration 0 and release planning have been done. They follow a routine cycle of activities:

1. Conduct Iteration 0, if needed, to set up for the iterations
2. Plan the Iteration, estimating and prioritizing the work to be done
3. Daily execution:
   - Daily review and adjustment
   - Develop and test continuously
   - Handle impediments as they arise
4. Demonstrate the product at the end of the iteration
5. Release to customers
6. Hold a retrospection on the iteration to learn how to improve
7. Incorporate new requirements
8. Prioritize and estimate work for the next iteration based on all requirements
9. Start again

Checklists

Checklists for Iterations begin on page 117.

"Iteration 0"

In most projects, before the first iteration of a release can be started, it is advisable to do some "pre-work": Analysis to set up stories for the first iteration and create the development environment including tool configurations. This is called "Iteration 0."

Iteration 0 is the work that is required to answer the questions:

- How do we know we are done planning?
- What documents do we have?
- What do we know?
- How do we know we are ready for the first iteration?
- What documents do we have?
- What do we know?

Iteration 0 may take *one to four weeks* to complete.

Checklists

A checklist for Iteration 0 is on page 115.

Prerequisites

The team understands the product vision and release plan.

Activities

In Iteration 0, the whole team does the following:

- Understand the goals of the release
- Agree to the general release plan
- Estimate story points for next iteration. Add required stories. Then let the team get started on the work of the first iteration
- Establish work environment
- Identify impediments to remove and analysis to conduct before work commences
- Identify lessons from previous releases or iterations that the team wants to incorporate into this release
- Review and revise the team's standards and practices

Iteration Planning

Iteration planning is done prior to the next iteration to prepare to get started. It is done by the whole team.

Checklist

A checklist for Iteration Planning is on page 117.

Prerequisites

Before an iteration can commence, the team must have populated the iteration backlog and finished "Iteration 0."

Activities

In iteration planning, the whole team does the following:

- Understand the goals of the iteration.
- Estimate story points for next iteration. Add required stories.
- Team commits to iteration plan.
- Update the work environment and project board.
- Identify impediments to remove and analysis to conduct to allow work to progress.
- Identify lessons from previous iterations that the team wants to incorporate into this release.
- Review and revise the team's standards and practices.
- Commit to demonstrating the product and reviewing lessons learned at the end of the iteration.

Add, Revise, Remove Requirements

As customers gain experience with the product, they learn more about what they need. They may development new requirements, may decide they no longer need something they had asked for, or may decide that something is more important or less important than before.

At the end of each iteration, based on what the customer has learned from seeing and using the product, the Business Product Owner (acting as advocate for the customer) can:

- Add new requirements
- Revise existing requirements
- Remove existing requirements

After these adjustments are made, the collection of requirements are then prioritized and used as input for the next iteration.

Issues and Considerations

Topic	Discussion
Lack of a known velocity	When the team's velocity is not known, populate the backlog using size rather than velocity.
Testing was not done	If testing has not been finished by the end of the iteration, then the team cannot know how much testing resource is available in the next iteration. Move testing early in the iteration.
Team composition	If the team composition changes, then they cannot know what their capacity (velocity) will be for the upcoming. Treat their velocity estimate as a guess, just like it was at the first iteration.
Interruptions	Establish a process to handle interruptions from management, other projects, etc. so that teams are not blind-sided by interruptions in the middle of an iteration. Perhaps schedule time at the end of the iteration for other work.
People are not available	If all team members are not available to plan together, the plan will almost certainly be incomplete, inadequate, and erroneous. Adjusting during execution is wasted effort.

Iteration Execution

The iteration is when the development work is done. The goal of the iteration is to complete the stories that have been agreed to and produce perfected and demonstrable product at the end.

During the iteration, the team is focuses on its work and should be relatively free of interruptions. They should be identifying impediments, adjust their work processes based on what they are learning.

At the end of the iteration, the team demonstrates the product, reflects on lessons learned and how they need to improve process, and is ready for the next iteration.

Checklists

A checklist for Iteration Execution is on page 119.

Prerequisites

Before an iteration can begin, the iteration backlog must be populated, the project board and team environment set up and cleaned up; and the team commits to the plan.

Activities

In iteration execution, the whole team does the following:

- Starting on stories
- Selecting tasks
- Updating the story board
- Completing stories
- The Daily Stand-up

Starting Stories

When selecting stories, it is important that the amount of work in process (WIP) matches the capacity of the team. Having many stories open runs the risk of having many stories open at the end of the iteration and reducing team efficiency. This should be avoided as much as possible. Having stories that are in process at the start of an iteration causes many problems – not the least of which is you are never sure how much time it will take to complete them.

When defining a story it is useful to ask the question, "How will I know I've done that?" Getting an answer to this question gives the developer an example of what the story needs to accomplish. This helps the developer check their understanding of the story. Not answering this question will often result in work being done that will later be discovered to not have been what the customer asked for.

If the Business Product Owner "does not have time" to answer this question, try to come up with the best answer you can and ask if that is what is meant. This requires much less time on their part but will still give an example.

Selecting Tasks

Focus on tasks that will help to complete open, in-work stories.

Updating the Story Board

Team members should themselves update the story board for the following events:

- Open story (typically indicated by having a task in process)
- Active task (typically indicated by having a task in a different place than it started)
- Impeded task (typically indicated by placing a sticker on it or moving to an impeded row or column)
- Completed story waiting for customer validation (indicated by moving it to a completed row or column)
- Done-Done story (indicated by moving it off the board)

Completing Stories

Stories are completed when they are done-done-done. This means:

- The system runs on the developer's computer as expected.
- The system is verified by running unit tests on a common machine.
- The system is validated as being of deliverable quality with functional tests.

Level	Test	Business PO	PO	Customer	QA	Dev
Business Value Chunk	Business Value / ROI realized	X				
Minimal Business Increment	Incremental Business Value / ROI realized		X			
Feature / Scenario	Completed and accepted		X			
User Story	Completed and accepted		X			
Customer / User	Customer acceptance test			X		
Functionality	Functional				X	
System	Integration					X
Component	Unit					X

> **Protect the team from wasted time**
>
> *Meetings are a big time waster in many companies. Team members should not have to attend meetings called outside the Scrum team unless the meeting is absolutely required (such as mandated employee management meetings). Meetings whose purpose is to report status can be eliminated through the proper use of visual controls and Gemba walks and through the end of iteration demonstrations.*
>
> *If a team needs to be represented at a meeting, the Scrum Master can attend.*

Activities of Business Delivery

The Daily Stand-Up (Daily Scrum)

The Daily Stand-up (also known as the "Daily Scrum") is the activity of the team to decide together each day what team members will focus on. It is quick, focused, and highly collaborative. Daily Planning is the responsibility of the whole team, each member talking with the rest of the team.

The purpose of the Daily Stand-up is to communicate progress, identify impediments, and create teamwork. It is *not to solve problems*. Problem-solving can be done afterwards, offline.

STOP Starting and START Finishing!

Checklists and Ground Rules

Checklists and ground rules for the Daily Stand-up is on page 122.

Prerequisites

Before the Daily Stand-up can begin, the status of stories, tasks, and impediments must be updated on the team's project board.

Activities

Daily planning involves a brief stand-up meeting involving the Scrum Master and the development team - technical and business. The Business Product Owner should try to attend whenever possible, even if it is a call-in.

The goal of the Daily Stand-up is to ensure consistent progress in completing the work required for the current iteration. Priority is on completing in-work stories rather than starting new stories.

The members agree to follow the simple rules of the Daily Stand-up.

Rule	Description
Be consistent	The Daily Stand-up meets every work day at the same time and in the same place.
Show courtesy	Team members show professional courtesy: show up on time, participate, and listen.

Rule	Description
Be brief	The Daily Stand-up typically lasts for 15 minutes. Standing up reinforces brevity. Extra discussions and problem-solving is conducted after the meeting, when there is more time.
Hold a team-wide conversation	The Daily Stand-up is for the team's benefit. Each team member is expected to speak and speaks to the whole team. The team is not reporting to the Scrum Master. The Scrum Master is there to help facilitate this conversation but not to lead the session.
Answer all three questions	Each team member answer three questions: • What have you done since last meeting? • What will you do before next meeting? • What is blocking or slowing you down? Team members can raise issues and obstacles but not propose solutions.
Review impediments	The Scrum Master reviews impediments and status.
Swarm	If you have the skills to help with a task and you do not have something else to do, you should volunteer to join the swarm.
Finish work	The priority is to complete stories that are in-work before starting new stories.

Tip. One of the most important jobs of the Scrum Master is to reflect on how the iteration is going. This includes looking for tasks that are taking too long.

Issues and Considerations

Topic	Discussion
What if the team falls behind?	The iteration plan is just that: a plan. When the team determines it is falling behind, it should not just hope it will catch up. It should let the Business Product Owner know so he/she can prepare to remove stories from the iteration if necessary. Be hesitant to open any new stories until those in process are done (a good practice anyway) in case not all of the stories get completed.
What if the team goes faster than expected?	If the team looks to complete its iteration ahead of time it should tell the Business Product Owner this so he/she can prepare to add some new stories to the iteration if they can be completed.
What if someone identifies an additional task in a story?	When an additional task is discovered, you must determine if this will add time to complete the story. Sometimes it does, sometimes it does not. If the additional effort is insignificant, there really isn't much you have to do. However, if the extra time now puts a story in jeopardy, it is important to ask the Business Product Owner if the story is still worth the extra cost.
How does the team make sure it is ready for the next iteration?	A little look ahead is required in order to be sure enough analysis has been done to be able to make good estimation and planning during the iteration planning day. A certain number of stories should be analyzed prior to the iteration to prime the pump.

Tip. The team may stop an iteration if they discover a significant barrier that cannot be resolved. The Business Product Owner and whole team must decide what to do and re-plan.

Product Demonstration

At the end of each iteration and at the end of the release, the product is demonstrated to the Business Product Owner and key stakeholders. This demonstration is not just a presentation but is a conversation between the participants about what was done and what to do next.

Ground Rules

Ground rules for the product demonstration are on page 120.

Prerequisite

The essential prerequisite is that everyone is available for the demonstration: the key stakeholders, the Business Product Owner, the Scrum Master, and the whole team.

Agenda

The Business Product Owner owns the demonstration and, often, the Scrum Master facilitates the meeting.

The demonstration is informal. It covers the following:

- The product vision, the release vision, and objectives for this iteration.
- What was done.
- What was not done, including surprises and challenges not yet overcome.
- The walkthrough of the product *as it is*.
- The team decides how to do demonstrate the product.

The walkthrough should cover more than simply the technical aspects: discuss what has been done with the people skills and training and the processes required to realize the feature. Adjustments to priorities (new, revised, removed) in order to achieve value in the next iteration.

> **Tip.** Sometimes, the customer may be satisfied enough with the product that she decides that it can be released as is so that the Business can begin using it. If everything else is ready, release it.

Issues and Considerations

Topic	Discussion
Useful stakeholders are not present	If stakeholders are not present, you cannot get proper feedback.
Managing expectations	When the product is only partially complete and not ready to be released, the stakeholders have to be very clear that it is not ready.
It is good enough	If the customer is satisfied enough, consider releasing it... *use it*!

Example: Web Self-Service

The Business identifies several actors that will use a new Web Self-Service Account Management system, including: Home Owners, Renters, and Business Owners. The Business decides that servicing Home Owners would be the most valuable and should be released first, followed by other actors.

Providing Account Management to Home Owners is part of the capability for Home Owner Self-Service.

Looking at this, the planning team composes Home Owner with the stories for Account Management to identify four scenarios: Home Owner Create Account, Home Owner Read Account, Home Owner Update Account, and Home Owner Delete Account (involving the support desk).

Looking at these, the Business decides that the first three are essential. Since Home Owner Delete Account requires help from the support desk, they judge that it is not essential for the computer system to do it automatically. They can continue using support desk personnel.

The Home Owner Create Account, Home Owner Read Account, and Home Owner Update Account comprise an MBI.

Elements: Vision for the Product

Developing the product vision is an activity of the Business Product Owner and key stakeholders to ensure that the essential message and direction that is driving the product and individual releases is clearly understood. They try to express this in one page that is memorable to everyone.

Each release should also have a release vision statement that helps the team understand the purpose of the release. The vision statement must be made visible and shared. It can be formatted in the form of a standard product vision chart or a *pro forma* press release.

Prerequisites

The Business has described new or expanded capabilities that are required by the product. This is what drives the vision.

Activities

The product vision and release vision should each be expressed in a one page summary which serves both to define the release goals and to entice customers with the most important benefits they will realize.

It completes the following template:

> **FOR** <target customer>
> **WHO** <statement of need>,
> **THE** <product name> **is a** <product category>
> **THAT** <product key benefit, compelling reason to buy>.
> **UNLIKE**< primary competitive alternative>,
> **OUR PRODUCT** <primary differentiation>

Elements

Creating the One Page Summary

Writing a "pro forma press release" is a fun, short exercise for creating the product vision and release vision. The Business Product Owner brings the essential description of goals and drivers and helps to guide the work.

The whole team engages in the conversation about what should go into a "press statement" to be released at the end of the effort.

The result is a deep understanding of the release by the Team. When finished, the One Page Summary should be converted into a poster and displayed in the team room.

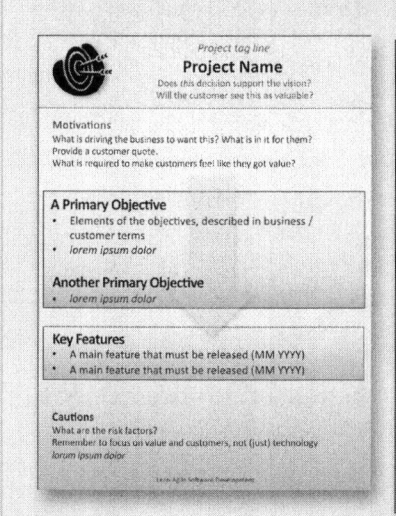

Project tag line
Project Name
Does this decision support the vision?
Will the customer see this as valuable?

Motivations
What is driving the business to want this? What is in it for them?
Provide a customer quote.
What is required to make customers feel like they got value?

A Primary Objective
- Elements of the objectives, described in business / customer terms
- lorem ipsum dolor

Another Primary Objective
- lorem ipsum dolor

Key Features
- A main feature that must be released (MM YYYY)
- A main feature that must be released (MM YYYY)

Cautions
What are the risk factors?
Remember to focus on value and customers, not (just) technology
lorem ipsum dolor

Press Release

For immediate release...

Your location, today's date

The *XYZ Company* announced today the successful release of the _____ product. This product provides _____

The customer for this product, _____, indicated in a recent interview that they selected this product due to the following key benefits:

 1
 2
 3

The customer also identified several features that they felt were particularly useful, including:

 1
 2
 3

XYZ President, _____, noted that the single most important benefit of this successful release was,
" _____ "

Elements: Features

When the release is presented to the Scrum team, each element of the Minimal Business Increment (MBI) is described as a "feature." Features are the essential planning units for releases, what the team uses to decide what is done and when. The team uses features to identify stories, which are the essential planning units for iterations. A feature is sized, prioritized, and placed on the product backlog. Features are traceable back to business value.

Essential Attributes of Features

The essential attributes of a feature are:

- **ID.** An identifier used to refer to the feature. Use an *intention revealing name*.
- **Business Value.** *(optional / helpful)* Indication of the value to the business. The Business decides on the scheme to use: a priority number, an estimate of ROI, an estimate of cost savings.
- **Description.** A few words to describe the feature. Here is a good template
    - Who or what area, including the scope and scale (for example: all managers, only active customers)
    - The capability / functionality needed
    - When is it needed (urgency)
    - How we know we have achieved the value
    - The method or approach for validation
    - Acceptance tests and criteria to validate value realization
- **Size.** The size, in "points" of the feature. Must be able to be completed within the release.

Elements: Stories

A story describes some aspect of a feature. Usually, there are many stories associated with a feature. Stories always trace back to business value through its associated Minimal Business Increment. Stories provide the ability to track work efforts in an iteration. A story might be a single scenario, an example of some behavior of the feature, or even a reminders or triggers to have a future conversation.

Stories are posted on the project team board and updated in the Daily Scrum.

Essential Attributes of Stories

The essential attributes of a story are:

- **ID.** An identifier used to refer to the story.
- **Description.** The intent of the story. It should be descriptive enough to allow a business analyst to write it several iterations after it was initially conceived. Here is a template:

    As a <user>, I want <capability> so that I get <business value or functionality>

- **Size.** Stories must be "right-sized" so that a teamlet can complete it within 1/3 to 1/2 of the iteration, ideally in one to several days. Can be expressed in "Business Value."
- **Validation Strategy.** Every story must have criteria so that the team can know, "Is this story done?"
- **Feature and Related Story.** Where did the story come from and what else is it related to?
- **Story Captain.** *(optional / helpful)* Add this to the card to indicate the team member responsible for shepherding the story to completion.

```
Story "as a <role>, I want <capability> so that I get <business value>"  ID: _____
Description: _____
            _____
            _____

Size (Points):  0  1  2  3  5  8  13  20  40  100  ?   Business Value: ____
Validation Strategy: How will I know I've done that?
            _____
            _____

Feature: _____   Related Stories: _____
```

Identifying Stories from Features

To identify stories from features, ask what it will take to do this:

- Define the functionality of the feature
- Build the functionality
- Release the functionality
- Support the functionality
- Use the functionality

This will lead to identifying stories such as the following:

Type	Description
Business Capability Story	A capability of the Business that must be developed for the feature to be developed. This would involve such as processes, documentation, training, or equipment.
Analysis Story	Research that needs to be done to describe the feature more fully, including studies, interviews, root cause analysis
User Story	How users interact with the feature, including requirements, test cases, high-level system design
Deployment Story	How the feature is introduced, supported, and marketed. Also known as a "Change Management" story. Includes the run book, configuration settings, environment specifications, application disaster recovery
Environment Story	Physical, logistical, or virtual configurations that must be established before the feature can be developed or deployed.

Elements: Tasks

A task work item communicates the need to do some work. Each Team member creates tasks unique to the role: developer tasks, testing tasks, impediment resolution tasks, etc. A task can also be used to suggest that exploratory testing be performed. A task can be used generically to assign work within the project.

Tasks are posted on the Project Team board and updated in the Daily Scrum.

Essential Attributes of Tasks

The essential attributes of a task are:

- **Description.** A concise overview of the task to be completed. The title should be descriptive enough to allow the team to understand what area of the product is affected and how it is affected.
- **Estimate.** Tasks must be "right-sized" so that a teamlet can complete it in one to several days.
- **Exit Criteria.** Every task must have an exit criteria so that the developer knows when the task is done.
- **Story.** Where did the task come from?

```
Task
Description: _____

_____
_____
Estimate: _____ hrs
Exit Criteria: How will I know I've done that?
_____
_____
Story: _____
```

Identifying Tasks from Stories

To identify tasks from stories, ask what it will take to do the following:

- Define the functionality of the story
- Build the functionality
- Release the functionality
- Support the functionality
- Use the functionality

This will lead to identifying tasks such as the following:

Type	Description
Design Task	Work that applies design patterns and coding principles and practices to specify code requirements.
Dev Task	Work that generates code from analysis tasks.
Testing Task	Work that defines validation and verification tests for stories and code.

Estimation in Agile: A Conversation

A primary benefit of estimation is the conversation that happens between appropriate stakeholders about the work to be done. Understanding is more important than accuracy: It validates that everyone understands what is required. Accuracy will come with experience.

Teams face a number of routine challenges to estimation including going into too much detail, making assumptions, doing design while estimating, and being reluctant to commit to an estimate. The Scrum Master must help the team get through these challenges.

Estimation follows the Plan-Do-Check-Act process from lean: Develop the estimate, do the work, check to see what was done and how much effort it took, and incorporate what was learned for the next iteration.

Estimation is done at every level of planning and involves everyone involved in that planning session. Estimates reflect the team's best guess at the moment. Early on, estimates may be +/- 100%. As learning grows, estimates become more accurate.

Estimating by points

Teams estimate in terms of points. A point is not a unit of time. It reflects an arbitrary judgment by the team about how "big" an item of work is. Team capacity describes how many points they can process. The point scheme used follows a sequence such as: 0, 1, 2, 3, 5, 8, 13, 20, 40, 100, 200, 400, ...

Estimating work to be done

Early-on in analysis, estimates are done for features. These are "coarse" estimates. As teams drill-down to stories and tasks, estimates get better: It is easier to estimate smaller things.

Estimating team capacity

Team capacity or "velocity" is the number of points that a team can process within the planning horizon (release, iteration). Estimating team capacity involves looking at historical trends: *this* team can handle *NN* points.

Issues and Considerations

Topic	Discussion
What if those doing the estimation only know some of the story?	It is not uncommon to have a team estimate a story where no one member knows the size of all aspects of the story. In other words, some teams may have members that understand how to write the code, others how to test it. In these situations, the team must figure out the best way to do estimation. A good approach is for those who are going to write the code to estimate that aspect of it and those who are going to test the code to estimate that aspect of it and then add the story points of each sub-estimate together to get a total story point for the story. For example, developers may say it will take 8 story points to develop a story and the testers say it will take another 5 to test it. The story estimate is 13 story points.
How we know we are done?	Teams will quickly discover that estimates without getting this question answered will usually be lower than what they would have estimated with this question answered. This is a very important question to answer - not just to better understand the story but also to help create better estimates.
The story will not be done soon	Estimates are not really needed for stories that won't be worked on for a while.

Team Estimation Game

The Team Estimation Game (created by Steve Bockman) helps teams size features and stories based on *relative* complexity. It works because people find it easier to compare the complexity of one feature or story with another even if they do not yet know all aspects of that story. This game is fast, easy, and fun. It helps people not to get bogged down into too many details, which is always a risk during estimation exercises. Relative ranking gives the team the information they need to decide what to work on and how much to commit to in current the planning horizon.

Remember, estimates represent our best guesses about the effort required based on what we currently know. As we gain more experience, we will have more information and can then refine the estimates.

One good way to think of feature or story complexity is the degree of connectedness of the feature or story. This can indicate how inter-connected it is with itself or the number of connections to other features or stories.

Traditionally, this game is played at the iteration planning session to size stories for the next iteration.

Another popular estimating technique is "Planning Poker." The Team Estimation Game is typically faster and easier to learn than Planning Poker and results in estimates of similar quality.

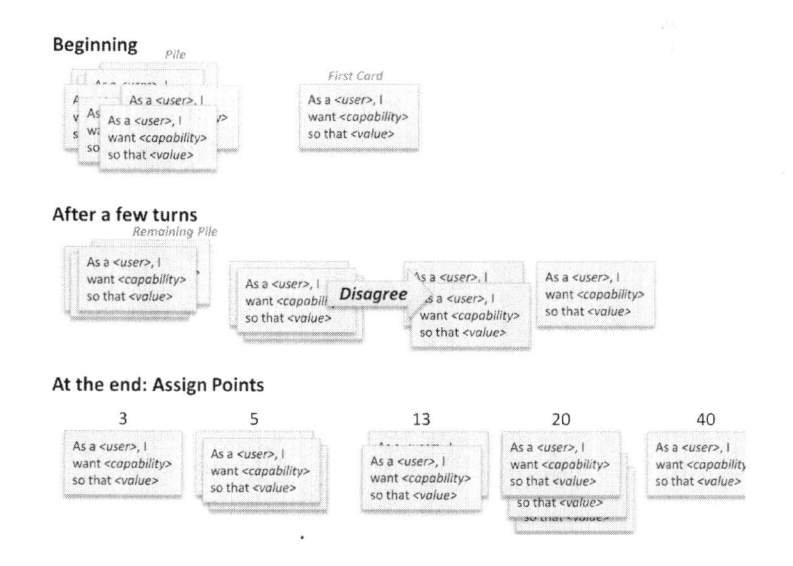

Beginning
Pile   *First Card*

After a few turns
Remaining Pile   Disagree

At the end: Assign Points

3   5   13   20   40

How to play the Team Estimation Game

Step	Do this...
Set up	Place story cards in a pile on the table. Select the top card in the pile and place it on the playing surface a foot or so away from the pile.
First play	A player takes the top card off the pile and places it somewhere on the playing surface indicating its size relative to the first card: to the left if it is easier, underneath it is the same size, to the right if it is more complex.
Relative estimation of the rest of the deck	Each person plays in turn doing one of the following: 1. Play the top card from the pile as described above 2. Move a card already on the playing surface, declaring disagreement about its relative size 3. Pass Play ends when there are no more cards in the pile and there are no more adjustments to be made. Note: During play, anyone may talk about why cards are being moved or about what they think about size of the stories. The goal is to get clarification and not to get too hung up on the exact sizes of the estimates. Remember that these are just estimates!
Assign points	The team works together to assign points to each stack to indicate the size (level of effort) of stories that are in that stack. Use the sequence: 0, 1, 2, 3, 5, 8, 13, 20, 40, 100, 200, 400, and 800. When done, write the assigned points on each card.

Part IV: Quality

Quality of product and process is the responsibility of *everyone* on the team. It is the result of discipline, communication, and the use of good engineering practices.

Continuous Improvement Before, During, and After

During each iteration, during every activity, the people doing work should help each other think about quality issues. Be explicit about what is required without trying to solve the problem (which is the responsibility of people who will receive stories or tasks to do). The TDM often gets involved in helping to think about issues of technical quality.

Improving quality involves discipline and mutual commitment. Lean techniques – such as retrospections, keeping a clean environment, testing, and a correct use of patterns and code quality – help, but only when the team uses them regularly.

One of the principles of Lean-Agile is to *create knowledge*. Scrum Masters should be sharing lessons (sanitized if need be) with fellow Scrum Masters through their community of practice. In fact, this should be part of the assessment of their performance.

> *Tip: STOP! If you do not pause to reflect, you will not learn. And if you do not learn, you cannot improve. And you will cease to serve your customer.*
>
> *Tip: STAY FRESH! At some point in every team's life, quality improvement becomes rote, stale. It ceases to be useful to the team. Whenever you sense this, it is your responsibility to point it out. The team must consider together how to shake things up, to see how they can make it relevant again.*

Retrospection After Every Iteration

Retrospection is the structured reflective practice to learn and improve based on what has already been done. The purpose of retrospection is to build team commitment and to transfer knowledge to the next iteration and to other teams.

Retrospections must be done at the end of every iteration. A briefer version, the "After Action Review," can be done at *any* time, whenever there is value for the team to stop and learn from what has been done and change while it still helps work.

Responsibility: Whole Team

Support: Scrum Master

Deliverable: 1-3 stories for the next iteration reflecting the "vital few" improvements to the process

Prerequisite

The essential prerequisite for a retrospection is that everyone who was involved in the iteration be available. The reason is that everyone has some viewpoint about what happened. If someone is missing, an insight will be lost.

Activities

At the end of each iteration, the whole team conducts a retrospection, facilitated by the Scrum Master. The key question is, "*If we could do it again, what would we keep doing and what would we improve?*"

To answer this, the team discusses the following:

- How well did we do in the iteration? Did we meet the iteration goals?
- How is the project progressing?
- What were your successes? (ask everyone)
- What processes / techniques would be good to use again?
- What were your challenges? What could have gone better? (ask everyone)
- What processes and techniques need to be improved?
- How is the team being? Are they raising impediments? Are they adjusting?
- What impediments are still present?
- On a scale of 1-10, what is the score of this iteration?

Guideline	Description
Facilitated	The Scrum Master facilitates the discussion and does not solve problems.
Whole team participation	Everyone who was involved in the iteration should be present at the retrospection.
	Everyone speaks because everyone has an insight that may help foster understanding.
Blame-free	The goal is process improvement, not blame. Be honest about what happened. Critiques are allowed without recrimination.
	Use the normal rules for brainstorming.
	Seek to uncover the unvarnished truth, what actually went on.
Vital Few	While a team may generate a lot of ideas, have the team pick a "vital few" that offer the greatest opportunity for near-term improvement. For each one, create a story and assign it to an iteration.
	A retrospection is successful if it generates 2-3 stories focused on process improvement.

Quality

Facilitator Notes

The retrospection facilitator is responsible for the following:

- Invite Scrum Masters from other projects to observe the retrospection so that they can carry lessons back to their teams, hearing subtleties not present in written reports.
- At the beginning, ask people to introduce themselves and their role (even teams that know each other)
- Begin by reviewing the objectives for the iteration
- Create an atmosphere of openness and don't be afraid to ask the unasked questions.
- Clarify the distinction between facts and opinions
- Ensure a blame free environment. Avoid a witch hunt!
- In a big meeting, ask someone to take detailed notes

At the end of the iteration, the facilitator should consider whether there are others in the organization who would find the lessons learned here useful to their situation.

Issues and Concerns

Topic	Discussion
Growing stale	At some point in a team's life, retrospections become rote, stale. It ceases to be seen as useful to the team. Whenever you sense this, it is your responsibility to point it out. The team must consider together how to shake things up, to see how they can make it relevant again. Consider looking more widely or deeply about processes, based on lean principles.
Too many suggestions	Focus on the vital few. Track these as stories for the next iteration so that the team can see value being produced.
Too few suggestions	Teams often think too small. They are constrained by assumptions about what they are allowed to do.
Complaining	It is common for improvement events to devolve into whining sessions, complaining without an intention to do better. The facilitator can allow a little time for this but then must take strong action to head this off.
Lack of a properly trained facilitator	Retrospections must be facilitated by a trained facilitator who is willing to lead without trying to solve problems.

Impediments to Progress and Quality

An impediment is any technical, personal, or organizational issue that inhibits progress being made on delivering product to the customer. Impediments are inevitable. It is always the responsibility of the team to expose impediments as they arise rather than burying them, hoping they will go away. The team is responsible for resolving impediments that lie within its "span of influence" (within which it can effect change). The Scrum Master, acting on behalf of the team, elevates all other impediments to management to be addressed. The Product Owner and other management agree to help.

Addressing Impediments

The team identifies impediments in the process of actually doing work. They should surface at least during the Daily Stand-up and the Iteration Retrospection. Whenever an impediment is identified, the Scrum Master writes a story for the impediment. The team prioritizes the impediment story and decides whether to resolve as a team or elevate.

Techniques for addressing impediments include:

- An After Action Review, targeting the impediment
- A quick improvement workshop (Kaizen) to address the impediment
- Launching a root cause analysis effort

> "Farming looks mighty easy when your plow is a pencil and you're a thousand miles from the corn field." – Dwight Eisenhower
>
> "Go see for yourself." The lean philosophy is based on management by fact, and the belief that facts exist where they are created, not far away from it. All improvement, whether it is technical innovation, process method redesign, or policy, must be based on the actual needs of the situation observed for oneself.

Impediments to Progress and Quality

Issues and Considerations

Topic	Discussion
Motivation	In the crush of work, teams are tempted to ignore impediments, or they identify too many impediments and never resolve anything. Both of these are almost always mistakes. The Scrum Master must push teams to stay disciplined.

Examples of Impediments

At this level...	Common impediments include...
Individual team member	Lack of training
	Environment
	Development environment is not available: hardware, software, tools
	Team infrastructure is not available: project board, tools
	Communication and customer
	Cannot talk with the person needed, when needed
	Analysts who do not know the customer, what they need, or who is most important. Do not know the voice of the customer
	Developers make promises inappropriately
	Not letting developers talk to customers unless analysts are present
	Cannot talk with customers
	Interruptions
	Having to multi-task
	Being interrupted by support issues
	Analysts working on too many projects at once, being asked questions from too many different developers

IV Quality

At this level...	Common impediments include...
Individual team	Project
	Lack of clear project vision
	Too much work in progress
	Unclear definition of done-done-done
	Not checking in code frequently
	Time
	People do not come to the Daily Stand-up on time
	People have different work hours, time zones, shifts
	Environment
	Not being co-located
	Not having an open Team Room
	Virtualization
	Interruptions
	Interruptions from management directives
	Management adds work to the backlog in the middle of an iteration
Cross-team	Lack of Value Stream Maps
	Lack of timely feedback from other teams
	Structural
	Team structure so end-to-end functionality is difficult
	Lack of relatively static teams (reforming after each iteration / release)
	Interfacing with non-Agile teams
Management	Product backlog not being updated
	Scrum Master not supported by management, team, or is ignored
	Code base
	Having to deal with old, legacy code that no one knows
	Poor quality code
	Lack of documentation (architectural, design)
	Having to attend meetings not related to development of the project

Impediments to Progress and Quality

At this level...	Common impediments include...
Testing and QA	Developer not wanting to talk with a tester, or a customer
	Developers not talking to QA until late in the process
	No testing environment, or running tests takes too long
	Cannot get access to data due to confidential nature
	Manual tests difficult to do and cannot do automated test
	No tools for testing databases pragmatically
	QA writing tests after code has been completed

Quality: Documentation, Standards

Aim for perfection! Improve constantly! Test Early and Test Often.

Highly performing organizations constantly think about what can go wrong and improve processes to address this. They welcome mistakes as opportunities to learn. They try to stay sensitive to conditions of the team. Quality involves making sure the team builds the right product and builds the product right.

- **Build the Right Product.** The close relationship between the Business and development and short iterations makes it easier to ensure that requirements are well understood. Customers have the opportunity to provide feedback early and often, when it is easiest to change.
- **Build the Product Right.** The goal of testing is not to discover defects but to prevent defects in the first place.

In Lean-Agile, quality is linked from features to stories to tasks. It is the responsibility of everyone on the team: Business Product Owner, business analyst, developer, and tester.

In software, quality involves having a language in which to discuss good practices and standards to assess how well code measures up. Testing is fundamental to the pursuit of perfection and is the responsibility of everyone on the team. It must be done as early in the process as possible and continuously.

The primary deliverable is a software product that the Business perceives as offering them value. Stay focused on this deliverable. In the course of this work, other artifacts are created. Use these but don't get distracted.

This document	Describes...
Acceptance tests, unit tests	What the features are supposed to do and the criteria for knowing when a work item is done
Team project board	The current status through various visual controls
Feature descriptions	The business value which the feature's functionality will achieve. This includes the Acceptance Tests that determine if that feature has been achieved.
Product vision	What the Business requires of the product, the charter for the project.

Documentation should always have a purpose. Never document simply for the sake of documenting. That is *non-value-added work* and that is waste.

The Lean-Agile Approach to Standards

The team must stay "mindful" of what the enterprise has learned: enterprise standards, best practices. As much as possible, begin with organizational coding standards. If it doesn't work, identify the standards as an impediment which then needs to be resolved by the team.

The Application Development Manager helps teams with understanding, using, modifying, and communicating changes to standards. The latter is important because discoveries by one team could well help other teams.

Patterns and Code Quality

Patterns (also known as "design patterns") focus on using existing quality solutions to solve recurring problems. As developers use patterns, both the team and the organization begin to experience improved quality:

- They incorporate better solutions than would otherwise have been thought of, because they are working with the experience of many developers.
- They gain a common vocabulary, which improves the quality of their conversation.
- They gain a deeper and more flexible understanding of object-oriented analysis and design.
- They have architectures that allow for more maintainable code. It is easier to change over time, is appropriately complex, and exhibits good code quality attributes.

Attributes of Code Quality

In both object-oriented and procedural code, assessing the quality of software can be difficult. Oftentimes, it involves a conversation among the development team as they think together about the code.

Quality	Description
Testable	Every story and feature has acceptance tests.
	Individual code units have unit tests, as needed.
	Test coverage is adequate.
Non-redundant	Objects/modules handle well-defined responsibilities.
	Follows once-and-only-once rule.
Encapsulated	Adequate encapsulation, hiding responsibilities at the data, implementation, type, design, and construction levels.
Cohesive	Strong class/module cohesion means that a class/module has a single responsibility and everything is about fulfilling it.
	Strong method/function cohesion means that each method/function is about fulfilling only one functional aspect of that responsibility.
	An object method implements a single behavior.
	Functions are related in a routine (or class).
	Strong cohesion relates to clarity and understanding.
Proper coupling	Classes/modules have strong internal relationships but small, direct, visible, and flexible relations to other classes/modules.
	Code is loosely interconnected.
Assertive	It is clear what each class/module is about and follows coding and naming standards as defined by the team. Objects tell each other what to perform.

> *The testing plan addresses all aspects of testing: Acceptance Testing, Automation, and Continuous Integration*

Quality: Testing

Testing involves more than simply discovering bugs. Testing helps discover the causes of errors and eliminating them. Testing helps make explicit the assumptions and requirements that customers have without getting overly technical. Testing ensures code integrity and compatibility with other code modules.

TO ENJOY THESE BENEFITS, testing should begin as early in the development process as possible.

Testing minimizes the risks caused by humans, machines, and environment.

Planning the testing strategy should also begin as early as possible. This requires careful consideration: what will be tested, how much can be automated and what must still be done manually, what the testing environment will include, who will do testing, what they need to know, and what standards they will follow.

Acceptance Testing

Acceptance testing helps the team answer the question, "Does the code do the right thing?" Acceptance testing is required by Lean-Agile.

Note: The issues that must be considered are described in *www.netobjectives.com/courses/lean-agile-testing*. For more information, see "The Role of Quality Assurance in Lean-Agile Software Development" in *Lean Software Development: Achieving Enterprise Agility* (Shalloway, Beaver, & Trott, 2009).

> *A system is accepted when it is "Four Done":*
>
> **Done:** *The system runs on the developer's computer as expected*
>
> **Done:** *The system is verified by running unit tests, code review, etc.*
>
> **Done:** *The system is validated as being of deliverable quality with functional tests*
>
> **Done:** *The system is ready for (pre-) production*

Acceptance tests come from the point of view of the user of the system. Factors to consider include:

- Up-front acceptance test specifications help the understanding of the requirement.
- Acceptance testing can be used to integrate development and test, increasing the efficiency of development.
- Acceptance testing can be used to improve the process being used to develop the system by looking at how to avoid errors from occurring in the first place.
- Automating acceptance tests can lower regression costs.

Acceptance tests are as much about the understanding and definition of the requirement as they are about validating that the system properly implements it. Ask the question, "How will I know I've done that?" This is something that opens the door for getting an answer in the form of an example - which can (should) be a test specification. This process of having an abstract statement (the requirement) and a corresponding concrete example (the test specification) improves the conversation between developers and analysts and testers and customers. Every feature and every story must have an acceptance test. These tests can involve both manual and automated methods. The outcome is to deliver fairly well-perfected code where defects have no serious consequences.

Automated Testing

Automated Testing is the use of software and hardware to verify and validate the behavior and qualities of the software under test. It involves software to control the execution of tests, to compare actual outcomes with predicted outcomes, to set up test preconditions, and to control reporting functions. Usually, test automation involves automating manual testing processes that are already being used.

Planning for test automation happens at the earliest stages and should include all levels of testing: unit, integration, system, system integration, and functional/acceptance.

As a rule,

All tests, including related setup, configuration, and evaluation steps, which can be automated, should be automated with priority given to aspects which are the most time consuming, error-prone, and tedious.

Common automated test frameworks include xUnit-based approaches (CPPUnit, JUnit, PyUnit), Boost test, and FIT.

When selecting an automated testing framework, consider the following:

- Adequate range of available interfaces: GUI-driven, an interface friendly to automation via scripts and command-lines
- Hardware-software dependencies and system requirements
- Procurement difficulty and cost
- Community support
- Feature mismatch between automatable and manual interfaces
- Performance
- Available API library
- Test data post-processing
- Report generation

Step	Description
Check out the software from the repository	Check out from the repository must be handled by scripting or other means in order to avoid error.
Build the software into the deployable or executable form	Compile (and link) the code in order to detect errors in grammar, etc. The automated tool should report errors to compile to the user in an effective manner, such as a scoreboard.
	Examine errors in the full context which was present when the errors were emitted from the compiler or linker.
	In the case of an automated build server, use a full-fledged, fully automated build system infrastructure.

Step	Description
Set up of test preconditions	Establish the dependencies required by the system or unit under test. This can involve many activities: • Simple initialization of a variable or instantiation of a class under test • Setup routine or table of a testing framework such as CPPUnit or FIT • Initialization of target nodes • Setting up environment variables • Connections to other systems, such as databases The degree to which developers can properly handle dependencies is the degree to which they can easily and properly test their software in a self-contained and automated fashion. Distributed, real-time and embedded (DRE) and network systems introduce unique complexities for establishing full testing preconditions, such as setting up COTS network test equipment via third party automation APIs as well as command line scripts, accessing and modifying a database of network topologies, and automated configuration of COTS network connectivity devices.
Deploy the software	This step may be straightforward when deploying on a single workstation or quite complex when deploying on DRE and network systems involving multiple levels of security.
Run the tests	If there are only a few tests, a script can automate running the tests. With a larger number of tests, use a testing framework to register and run the suite of tests. Use "Test Runners" or "Test Conductors" to orchestrate the control and actions of many pieces of test equipment, test software, and test result display to fully execute a set of complete functional tests.

Step	Description
Control software and resources that the Site Unit Test requires	Whether testing a module in isolation testing the module with the actual collaborators, the test strategy needs to control the other software entities: their lifetime (creation, initialization, configuration, destruction) and their behavior during the tests.
	A simple technique is to use a mock object that is part of a runtime polymorphic hierarchy and is hidden behind an interface. This approach allows for the substitution of the mock or the real entity depending on which kind of test is being run. More complex domains will require more sophisticated approaches; for example, controlling the network traffic entering nodes of a large, heterogeneous, mobile *ad hoc* network.
Retrieve and display the results	The testing framework should enable viewing test results, ideally in a simple scoreboard display or other standardized report. As far as possible, they should help developers find which tests failed, which errors occurred, and where.
Notify team members	Rather than requiring the team to monitor the test, the testing framework should actively notify the team about results so that they can isolate and fix problems quickly and effectively.
Information statistical analysis	The test strategy should be informed by statistical correlations of test results and other metrics that show how testing is progressing over time. For example, code coverage over the course of a week or growth rate of the number of test cases.

Quality

Continuous Integration

Continuous integration is the automation of the build and test process, all triggered automatically when code is checked into the Source Control Management (SCM) repository. It offers the team rapid feedback on the integrity and quality of the code.

The basic aspects of CI include:

- Automated, scripted builds and testing
- Using a dedicated CI server / build farm
- Triggering the CI server on check-in
- Building as frequently as possible
- Build the entire system
- Run automated test suites
- Notify team when build breaks or a test fails

Advanced aspects include:

- Static analysis
- Dynamic analysis
- Performance / Load testing
- Database testing
- Automated document generation
- Build deployment packages
- Deploying to live test environments

In order to configure the SCM system to trigger the CI server on a check-in, developers usually require administrative rights to the SCM server. To support this, the SCM server provides "hooks" that allow operations to occur before and after a successful check-in.

For example, if a project is using Subversion, developers configure the post-commit hook to trigger the CI server. The CI server receives information on the SCM repository such as branch, revision, changed files, and comments the developer made regarding the check-in. Based on this information, the build master delegates jobs to a series of predefined build slaves.

Automated, scripted builds and testing

Many systems involve modules that come from a variety of sources: subcontractors, open source, third party closed source libraries, and in-house code. Each of these individual software modules have their own set of build procedures; their own set of dependencies on other software modules. Keeping track of these dependencies is difficult. Not having the ability to track these inter-package dependencies explicitly is a recipe for disaster.

Handling inter-package dependencies should be implemented in an automated way rather than relying on a series of manual steps. Such an approach should support the following:

- Building from clean source
- Ability to clean directories safely, in order to restore the directory to a pristine state, removing executables, etc.
- Safe and fast incremental builds
- Options for compiling with different compilers and operating systems to the extent possible

Using a dedicated CI server / build farm

The build farm is comprised of one or more computers ("build slaves") dedicated to carrying out any instructions that the continuous integration server (the "build master") desires. These instructions involve a set of executable procedures to update from source control, build and test the software, and finally retrieve and return results. Results are then radiated out to team members, for example using an HTTP accessible scoreboard, e-mail, RSS, or SMS text message.

Using multiple build slaves reduces the time it takes to detect defects. For example, if a project has five components and six build slaves are available, dedicate five build slaves to unit testing a single component and dedicate one build slave to running integration and system level tests. This gives fast feedback to developers who need to know about failures in unit tests while running the (lengthy, and potentially CPU intensive) integration tests on another build slave.

Triggering CI server on check-in

The continuous integration infrastructure should be transparent to developers: Developers check code into the Source Control Management (SCM) repository and then the CI system automatically commands the build farm to carry out the appropriate jobs. If any of the tests break, the CI reports which check-in caused the problem. The team can find and fix the problem quickly or at least back out changes from source control.

Build as frequently as possible

In Lean-Agile, build and test is done frequently, ideally on every check-in to detect any errors or any degradation in performance. This minimizes the number of changes to be built and tested at any one time and this reduces the risk of introducing errors.

Build the entire system

Developers spend less time doing rote work (e.g. setting up complex test equipment, configuring XML files for test scenarios) and more time solving complex domain-specific problems.

> *Tip. If tests take a long time to run, say over 16 hours, it is better to dedicate hardware to running these tests continuously than run them less frequently. When teams do not have tests running continuously, developers will (inadvertently) break these tests.*

Is it best to build the entire system from scratch or to automate an incremental build? The trade-off is between speed of build (in order to get rapid feedback) and avoiding introduction of errors and correctness of the build. If the difference in time between a clean build and incremental build is large enough, automate both an incremental build and a clean build: use the former for rapid feedback and the latter to ensure correctness. Doing this task can easily be implemented with multiple build slaves.

Run automated test suites

Generally, the more tests that are automated, the less time the development team has to spend running manual tests and the more time they can spend implementing new features, automating tests, and solving domain-specific problems.

Running the software and retrieving results may not always be a simple task, especially if the team is targeting an embedded platform. These embedded devices often lack programmable facilities for deploying software and retrieving results. For many projects and teams in the embedded software arena, this prevents them from wanting to use Continuous Integration. The investment in creating the CI server will pay off in the long run.

If the targeted platform cannot be automated, try breaking the dependencies on this target and run the software "off target." This requires an explicit, formal commitment to go back to the objective target eventually.

Getting rapid results back to developers is absolutely priceless and dramatically increases the velocity of the development team.

Notify team(s) when build breaks or a test fails

After running the automated test suite, the results are radiated out to the team, posted to the CI server's scoreboard. The CI server software provides facilities for sending this information out, for example via e-mail, RSS, or even SMS text message. Anyone who has both the authorization and desire to see the information should be able to have access to it.

If a test happens to break, the person who checked-in the failing piece of code is responsible to fix the problem. If that code conflicted with someone else's code, both parties must resolve the conflict.

Beyond Basic CI

Continuous Integration can involve more than merely integrating changes and automating tests. These extensions include new types of tests, performing post-processing on the build and test artifacts, and anything that lends itself to being run in an automated fashion. A few areas for additional efforts appear below in no particular order.

Advanced Topic	Description
Static analysis	Analysis which is performed on the source code of a program in contrast to testing which is performed by actually executing the program.
	Examples: Checking for uninitialized variables, checking the code against coding style standards, checking for unreachable code, type safety, and checking for potential memory leaks.
Dynamic analysis	Analysis performed during the execution of a program. Examples: Memory usage analysis, code coverage analysis, performance profiling.
	Examples of tools include gprof (performance profiling), gcov/lcov (code coverage), and valgrind (memory analysis)
Performance/ Load testing	For many software applications, a certain level of system performance is an important non-functional requirement. The customer's requirements may even dictate the use of acceptance testing to verify a certain threshold of system performance has been met.
	When it comes to CI, if this type of testing involves third-party equipment, special consideration should be given at the earliest stages of test planning to ensure that the third-party solutions can, indeed, be automated.

Advanced Topic	Description
Database testing	Many systems heavily depend on a database. They need to minimize the impact of change and ensure integrity. In such cases, when it comes to CI, consider the following: • The team's database administrator should be part of a tightly knit cross-functional team so that everyone involved with the database schema is in close contact. • Use a per-developer, local version of the database schema and sample data in addition to a "shared master" version which committed changes are integrated against. • Use an evolutionary database design approach.
Automated documentation generation	The most common form of automatic documentation generation involves documenting the program's API. This works by parsing special symbols in some of the comments of a program's source code.
Build deployment packages	Automatic generation of deployment packages ensures confidence in "release at any time." This process varies from team to team and could include packaging the software binaries resulting from the CI build stage together with the system's documentation. Depending on requirements for deliverables, this may or may not include the program source code. For other projects, this may include creating an executable package to facilitate the installation process.

Advanced Topic	Description
Deploy to live test environment	Recreate the production environment as much as possible. If a live test environment is not readily available, this is difficult.
	Example. The need for remote management of the power and network connectivity of the system under test could create a requirement for a remotely accessible power distribution unit and remote control of the network.
	Resource management is another aspect. Examples include: hardware targets, network connections, test equipment, and servers. Represent these resources in a database which tracks their status in terms of configuration state and availability for use in a test. A server uses the information in this database to allocate resources to tests, as well as report status on tests in progress and/or the state of individual devices.
	Automated asset allocation is important when a team needs concurrent testing of different system-level configurations which use many of the same types of underlying resources.
	Example. A network of embedded targets can comprise the system under test. Testing this system involves multiple different tests running simultaneously. This requires automatic resource allocation. Identify as many of the requirements for full test-environment automation as possible and cast the entities involved as resources with an eye for how they could be managed in an automated way.

Here are references for more information:

- Static Analysis: *en.wikipedia.org/wiki/List_of_tools_for_static_code_analysis*
- Dynamic Analysis: *en.wikipedia.org/wiki/Dynamic_code_analysis*
- Database Testing:
  The Database Agility course describes a test-driven approach to evolutionary database design; see *www.netobjectives.com/courses/test-driven-development-database-boot-camp* and *martinfowler.com/articles/evodb.html*
- Automated Document Generation: *en.wikipedia.org/wiki/Comparison_of_documentation_generators*
- Build Deployment Packages:
  en.wikipedia.org/wiki/Installer#Installer and *en.wikipedia.org/wiki/Package_management_system#Common_package_management_systems_and_formats*

> **Unit Testing Frameworks and their Relation to Automation**
>
> *People often think that xUnit is for unit testing and FIT (or an equivalent) is for acceptance testing. In fact, developers can use both tools equally for these purposes.*
>
> *The xUnit architecture relieves the developer from many of the repetitive tasks associated with unit testing: setting up preconditions, registering tests, executing tests, tearing down tests, defining test suites, presenting test results, reporting state changes, and making assertions.*
>
> *FIT allows for acceptance tests to be specified by a non-programmer and even run by a non-programmer once the fixtures are in place. FIT provides powerful facilities for creating, setting up, tearing down, running, reporting, and displaying tests. It also provides facilities for full automation via command line scripting.*
>
> *JUnit: A Cook's Tour (Sourceforge 2009) and xUnit Test Patterns (Meszaros 2007) contain excellent discussions of the patterns and pattern language that underlie the xUnit family of frameworks.*

Communicating Information

Do everything possible to ensure clear communication: between team members who are doing work; between customers, stakeholders, and the whole team; between teams within the organization. This lies at the heart of much of Lean-Agile practices.

The Scrum Team Room

Whenever possible, the development team should be co-located. High bandwidth communication is crucial for efficiency and speed and not making mistakes. Ideally team members are located within 100 feet of each other.

Scrum teams should have a room dedicated to the team. All relevant information, especially the Project Board, is posted and maintained in this room. The Daily Scrum takes place in this room. Teams use the room for meetings and teamlets may also use it for swarming work. Here are important considerations:

- A sufficient number of developer cubes plus some additional work stations
- Direct, open visibility between all team members
- Gathering points (e.g. small round tables) in the middle of the room allow team members to talk together
- Walls covered with white boards
- A dedicated board for the Information Radiator

Information Radiators & the Team Project Board

The information radiator is a type of visual control that provides visibility about the project to everyone who wants to know about it. Information radiators are always up to date. Whenever management and stakeholders need to know team performance and progress on product development, they visit the workplace and review the team's project board. Visitors should be able to identify projects and who to talk to for more information. At a glance, the team should be able to know what is going on and who is doing what and decide what needs to be done next. This requires making information visible in one, predictable place.

While everyone on the team is responsible for keeping the project board up to date, the Scrum Master does most of this work.

Communicating Information

For teams that are not co-located, they must have one, common project portal for all of the information. Often, teams will maintain two sets of boards: a physical board in the team room(s) and the project portal.

Essential elements of the team's project board include:

- The product vision / release vision poster
- Contact information, especially the Business Product Owner, and Scrum Master
- Glossary of terms
- The team's rules
- The Agile scorecard (burn-down and burn-up charts)
- The list of impediments
- The product backlog stories
- The iteration stories and tasks

Here is an example layout of a physical team project board.

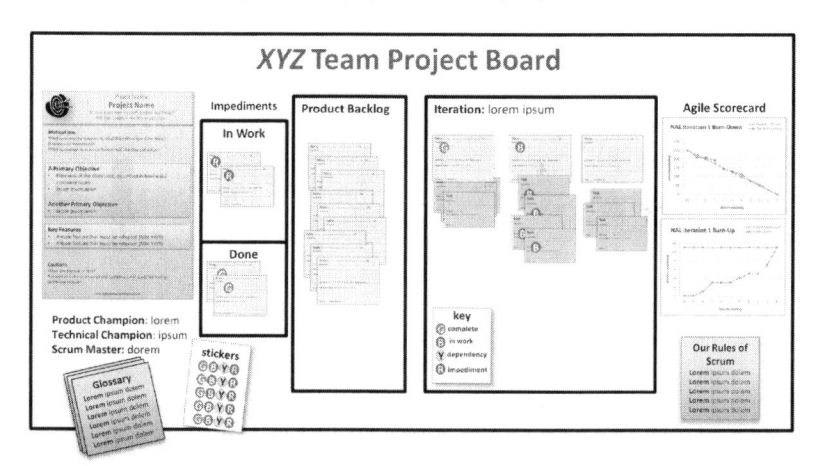

The Agile Scorecard shows the story point burn-down and story completion burn-up charts for the iteration. At the end of the iteration, the lines should join.

109

Here is an example of an Agile Scorecard showing burn-down and burn-up.

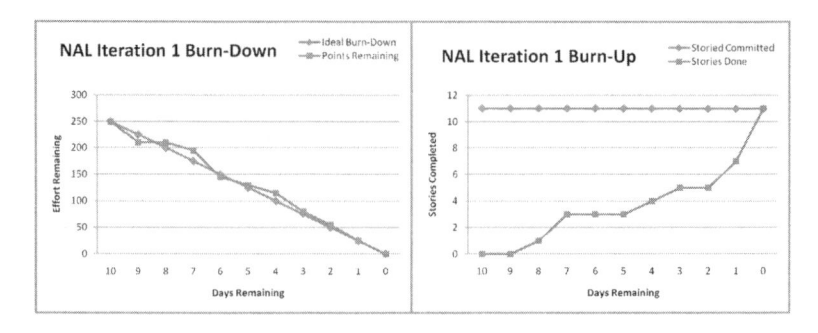

NAL Iteration 1 Burn-Down (Ideal Burn-Down, Points Remaining)

NAL Iteration 1 Burn-Up (Storied Committed, Stories Done)

Considerations for Non-Co-Location

Communication is 7% written, 17% verbal, and 76% body language. When teams are not co-located, they lose up to 76% of their ability to communicate effectively. Unfortunately, co-location is not always possible. For example, a project may require using multiple, large development teams or teams may be distributed across multiple sites (local or across countries). This requires more effort by the Scrum Master and the team.

Pay attention to the following:

- Project team members should be grouped together in their physical locations as much as possible.
- To maximize communication bandwidth, use dedicated broadband communication as much as possible: audio, video, story-boarding.
- Use an Agile tracking tool for managing product backlog, impediment backlog, stories, tasks, and tests so that teams can collaborate asynchronously.
- The team has to be very disciplined in creating and using artifacts, code, and discussion repositories.
- Teams must define usage criteria for code and tools and must be consistent across all teams on the project.
- Teams must coordinate their schedules for collaboration, planning, Daily Stand-ups, and reviews.

Part V: Checklists for Lean-Agile

Checklists are useful reminders for teams to ensure they are considering all of the practices of Lean-Agile.

Starting an Agile Team Checklist

✓	Activity	Description
	Select the right people	Select the Business Product Owner. *This is a crucial decision.* Select the ADM and TDM. Select the Scrum Master. Select candidates for the team.
	Select the right project	Select candidate projects that have a good likelihood for visibility and success.
	Train the people	Allow enough time to train the team with the skills they require. Allow enough time to give developers specialized training • Acceptance Test-Driven Development • Test-Driven Development • Object-orientation and design patterns • Automated acceptance testing
	Build the physical environment	Establish and/or build out the physical environment for the team: • The Scrum Team room • Visual controls, team's project board Decide how to handle teams that are not co-located Agile team.
	Tools	Acquire, build-out, and configure tools and environments for the team: • Agile life-cycle management tool • Source control management • Continuous integration server • Test frameworks
	Plan for assessment	Develop a plan for assessing maturity of Scrum Teams. Develop a plan for assessing Agile project success (metrics).

111

Checklists for Lean-Agile

✓	Activity	Description
	Transition plan	Develop a plan for making the transition.
		Gather Scrum Masters into a community of practice.
		Gather Business Product Owners into a community of practice.

Business Discovery: Priority Checklist

✓	Activity	Description
	Vision	Business Product Owner understands the product vision.
	Priority	Business value criteria established
		Chunks of work prioritized relative to other approved items
		Chunks of work sequenced and approved to continue

Business Discovery: Planning Checklist

✓	Activity	Description
	MBI	MBIs defined
		MBIs prioritized and sequenced based on ROI (relative to other approved items)
		Technical feasibility assessed
		Sized (using t-shirt sizing)
		Approved to continue
	Architecture	Architectural goals and approach identified, visible
		Dependencies and risks identified, visible
	Roles	Product Management assigned
		Business SMEs identified
		Product Owner assigned (if possible)

Business Discovery: Staging Checklist

✓	Activity	Description
	MBI	
	Features	Features defined and sequenced based on business value
		Acceptance criteria defined
		Sized (using t-shirt sizing)
		Approved to continue
		Sequenced based on ROI
	Business Backlog	Business backlog established

Business Discovery: Ready-to-Pull Checklist

✓	Activity	Description
	Business Backlog	Business backlog defined
	Roles	

Checklists for Lean-Agile

Business Delivery: Iteration 0 Checklist

✓	Activity	Description
	Roles	ADM and TEM assigned
	Team	Team identified with all of the needed roles, dedicated to the release, and co-located as much as possible.
	Resources	Required resources and environment available
	Vision	Business Product Owner has prepared the release vision.
		The team understands and agrees to the vision, drivers, and expected outcomes for the release.
	MBI	MBIs introduced and reviewed with team
		Refine description, scope, validation, and done criteria
		Identify risks, issues, dependencies, uncertainties, and known impediments
		Rank, prioritize, and mitigate with stories
	Technology components	Identify technology components
		Input technology features and/or stories into team backlogs
	Artifacts	Identify product documentation that must be maintained
	Architecture	Review and define the high level architecture and design
	Team environment	Review and define workflow
		Review and define visual board
		Review policies for stories
		Establish logistics for meetings, locations, times, frequency, participants
	Feature sequence	Finalize feature sequence by business value and technical dependencies

115

V Checklists for Lean-Agile

✓	Activity	Description
	Story estimation	Review and define complexity factors that will be used for relative sizing in story points
		Stories decomposed and right-sized
		• Analysis stories if needed
		• User stories for first feature
		• Validation criteria for stories are understood
		Stories are estimated for first few iterations' work
	Iteration backlog	Iteration length is set.
		Iteration backlog established, populated, and visible.
		Input items into appropriate tools
		Team has committed to Iteration 1 plan.
	Continuous improvement	Intentionally incorporate lessons learned from previous releases.
	Testing agreements	Testing approach (Unit, Integration, Acceptance) is committed to and visible.
	Reporting and metrics	Book of Work program backlog
		Topline
		Feature burn-up

Business Delivery: Iteration Planning Checklist

✓	Activity	Description
	Vision	The team understands and agrees to the iteration vision.
	Learning from the past	For the kind of work we are going to do in this iteration, are there lessons learned or good practices that we incorporate from anyone on this team, on other teams in the organization, or in the literature?.
		Invite a "Second Set of Eyes" from outside to think through the plan.
	Story estimation for iteration	Initial estimate of story points for iteration.
		Stories decomposed and right-sized.
		Validation criteria for stories are understood.
		Stories are estimated for first few iterations' work.
	Dependencies and risks	Dependencies and risks identified as stories.
		Impediments identified, posted on team board, assigned.
	Iteration backlog	Stories are assigned to the iteration backlog.
		Tasks are identified for the stories in the iteration backlog.
		Team has committed to iteration plan.
	Commitments	Commit to demonstrating the product to key stakeholders at the end of the iteration.
		Commit to conducting a retrospection at the end of the iteration.
		Commit to conducting After Action Reviews as appropriate *during* the iteration.
	Testing agreements	Definition of Done established and documented (unit, integration, acceptance).

V Checklists for Lean-Agile

✓	Activity	Description
	Team	The team is staffed with all of the needed roles, dedicated to the release, and co-located as much as possible.
		Team has received required training: Lean-Agile software development, Test-Driven Development, Engineering practices.
		Artifacts and Deliverables determined (and visible).
	Team environment	Tools for testing, coding, integrating, building selected and installed.
		Established logistics for Daily Stand-Up (time, location, conference call information, portal, etc.).
		Agreed to the ground rules for team life.
		Organized (and clean up) the team work space: physical, communication, collaboration.
		Set up the team's project board.
		Established the test and build environment.

Checklists for Lean-Agile

Business Delivery: Iteration Execution

✓	Activity	Description
	Input tests / code until tests pass	Unit test has been completed and documented. Defects clearly defined, resolved and tested.
	Run functional tests	Functional tests have been completed, documented. Defects clearly defined, resolved, and tested.
	Run user tests	User tests have been completed, documented. Defects clearly defined, resolved, and tested.
	Update business processes and conduct training	All impacted business processes have been assessed. Appropriate changes have been made. Training has been developed and executed. Participants have been certified.
	Business PO / Analyst acceptance	BPO or analyst has formally validated that the story has been completed and is ready to be promoted to production to support the MBI

Business Delivery: Product Demonstration

Rule	Description
Demonstrate the product as it is	The demonstration is always done with the product as it is. Never use slides or other artificial "demo-ware." This is possible because the product should always be tested (perfected), usable, and completed by the end of the iteration.
It is a conversation	The demonstration is a conversation between the whole team, the Business Product Owners, and the stakeholders. They collaborate on what has been done, what has not been done or could not be done, and what the current needs are. It is not a presentation.
Blame-free environment	The team has done what it could do and is forthright about what it could not do. It is not a time to assess blame but to describe the facts about what was really done.
Everyone is present	Everyone has a viewpoint to share. Many ears help maximize communication. As much as possible, avoid redundancy (waste) in meetings.

Business Delivery: Release Checklist

✓	Activity	Description
	Vision	Business Product Owner has prepared the product vision and release vision.
	Essential stories	Release planning team has identified the essential stories for the next release.
	Release plan	Release planning team has populated releases and prioritized MBIs.
	Demonstration	Team demonstrated product to key stakeholders.
	Retrospection	Retrospection of the release is planned. After Action Reviews conducted as appropriate *during* the release plan.

Business Delivery: Iteration Implementation

✓	Activity	Description
	Ready for Production	Code is ready to be released
	Promote	Release has been successfully executed
	Value extracted from feature for Business	Feature has been successfully executed and the Business value has been achieved
	Retrospection	Meeting has been conducted to evaluate success of work effort and ways to improve the process

Daily Stand-Up

✓	Activity	Description
	Setup	Status of Stories and tasks have been updated before the Daily Stand-up.
	Conduct	Team members showed up on time.
		All team members were present.
		Team members talked with each other.
		Problem-solving and side conversations were kept to a minimum.
	Team and environment	Team work space is clean: physical, communication, collaboration.
		Team's project board up to date.
	Impediments	Progress on resolving impediments was reported.

Daily Stand-up Ground Rules

Rule	Description
Be consistent	The Daily Stand-up meets every work day at the same time and in the same place.
Show courtesy	Team members show professional courtesy: show up on time, participate, and listen.
Be brief	The Daily Stand-up typically lasts for 15 minutes. Standing up reinforces brevity.
	Extra discussions and problem-solving is conducted after the meeting, when there is more time.
Hold a team-wide conversation	The Daily Stand-up is for the team's benefit.
	Each team member is expected to speak and speaks to the whole team.
	The team is not reporting to the Scrum Master. The Scrum Master is there to help facilitate this conversation but not to lead the session.

Checklists for Lean-Agile

Rule	Description
Answer all three questions	Each team member answer three questions: • What have you done since the last meeting? • What will you do before next meeting? • What is blocking or slowing you down? Team members can raise issues and obstacles but not propose solutions.
Review impediments	The Scrum Master reviews Impediments and status.
Swarm	If you have the skills to help with a task and you do not have something else to do, you should volunteer to join the swarm.
Finish work	The priority is to complete stories that are in-work before starting new stories.

Schedule: Various Scrums

Scope of Scrum	Schedule	Issues
Scrum Masters: *Team Progress*	Daily	Issues and impediments Improvements and adjustments Planning and coordination
TDMs: *Delivery Progress*	Daily	Impediments Readiness and coordination Skills, resources, and capacity
BPO and Analysts: *Value Progress*	Daily	Feature and story preparation Define Business requirements Validate completed stories
Lead TDM, BPO, Scrum Master: *Daily Value Delivery*	MWF	Progress and scope adjustments Priority, sequence, capacity impacts

Checklists for Lean-Agile V

Schedule: Iteration Planning

First Day of Iteration Planning: Morning

Time	Members	Topics
30 min	TDM, Lead, Tem Members, Scrum Masters	Close out previous iteration • Close completed stories • Split partially complete update sizes • Update Feature Burn up
30 min	BPO, PO, TDM, Lead	Show product completed • Can be tests passed • Improvements – practices and/or techniques
30 min	BPO, PO, TDM, ADM, Lead, Scrum Master	Conduct Group Retrospective • Identify any changes / improvements for next iteration • Commit to at least 3 small incremental improvements
3 hours	BPO, PO, TDM, ADM, Leads, Scrum Masters	Iteration Program Backlog • Define next Iteration's Goal • Load outstanding right sized stories from previous Iteration (if still priority) • Present user stories in priority sequence to accomplish the Iteration Goal • Validate / clarify / understand the User story • Define Right sized stories to accomplish the user stories • Size & sequence the right-sized stories and commit them to the iteration backlog • Update Planned next Iteration Backlog

125

First Day of Iteration Planning: Afternoon

Time	Members	Topics
2 hours	TDM, ADM, Leads, Team Members, Scrum Masters	Iteration Team Backlog • Present the Iteration Goal • Review the right sized stories planned • Decompose further into technical stories (right sized), update size in story points. • Define tasks, estimate in hours • Commit • Update Iteration Team Backlog(s)

Second Day of Iteration Planning: Morning

Time	Members	Topics
2 hours	TDM, ADM, Lead, Scrum Master	Finalize Team Backlogs • Review committed team stories and identify any adjustments, dependencies, and issues • Reorder (if necessary) and/or add stories to mitigate • Update consolidated committed Team Backlogs • Team members start work

1 hour	BPO, PO, TDM, Leads. Scrum Masters (optional)	Complete Iteration Planning
		• Review finalized committed Iteration Program Backlog with BPO/SMEs
		• Update Iteration Program Backlog
		  – Starting top line (in story points)
		  – Task burn down (in hours)

Part VI: Resources

Lean-Agile is a rich domain. This pocket guide touches on a minimum set of concepts you need to remember. This part contains a number of resources to further your knowledge and skills.

A Lean-Agile Glossary

Communication is essential to the team's quality improvement. It describes the shared understanding of vocabulary used by the team. It lists terms that are specific to the current product development effort. It is constantly updated as new terms are discovered or invented to ensure a common language.

The terms in this Lean-Agile glossary are in common usage. They have been gathered from a number of sources including books, journals, web sites, and field experts.

Term	Description
5S	A basic lean concept that helps to create an efficient and effective environment for work: "A place for everything and everything in its place." Derived from five Japanese terms beginning with "s" used to create a workplace suited for visual control and lean production. 1) *Seiri* means to separate needed tools, parts and instructions from unneeded materials and to remove the unneeded ones. 2) *Seiton* means to neatly arrange and identify parts and tools for ease of use. 3) *Seiso* means to conduct a cleanup campaign. 4) *Seiketsu* means to conduct seiri, seiton and seiso daily to maintain a workplace in perfect condition. 5) *Shitsuke* means to form the habit of always following the first four S's. Good references for 5S are in most Lean books.
Acceptance Testing	Formal testing conducted to determine whether or not a system satisfies its acceptance criteria and to enable the customer to determine whether or not to accept the system.

Resources

Term	Description
Agile	Agile software development is a conceptual framework for undertaking software engineering projects that embraces and promotes evolutionary change throughout the entire life-cycle of the project. Scrum and XP are two software development methods based on the Agile framework. See also: Scrum, XP.
Backlog	The set of stories that are not yet done.
Benchmarking	A technique in which an organization measures its performance against the performance of best-in-class organizations, determines how those organizations achieved their performance levels, and then uses the information to improve its own performance. Organizations may be internal or external to the company and may be in a different type of industry. Subjects that can be bench-marked include strategies, operations and processes. Sometimes, third-party organizations, such as the American Productivity and Quality Center, are used to conduct the benchmarks to protect confidentiality.
Best Practice	A superior method or innovative practice that contributes to the improved performance of an organization, usually recognized as "best" by other peers for given contexts. It is often best to focus on "good" practices.
Brainstorming	A technique teams use to generate ideas on a particular subject. Each person on the team is asked to think creatively and write down as many ideas as possible. The ideas are reviewed after the brainstorming session. Commonly used in kaizens. Most books on facilitation discuss the many approaches to brainstorming.
Build Verification Test	A group of tests to determine the health of a build at a high level. Typically, these tests exercise the core functionality of code and help determine whether further testing is warranted. These are also known as "smoke tests."
Burn-down	The rate (story points/day) at which stories or tasks are being completed.
Burn-up	The rate (story points/day) at which business value is growing.

Lean-Agile Glossary

Term	Description
Business	"The Business" is a short-hand for someone from another part of the organization who is not part of the technical development team but has expertise or responsibility for what the organization does.
Business Capability	The combination of people and people skills, processes, and technologies required in order to equip the Business / customer to use a product. All three are required, for "until the Business can begin to use a product, the product merely represents potential."
Business Value	What management is willing to pay for. It is a way to identify the value of "work" or a story.
Code Analysis	The process of checking that code conforms to design guidelines, looking for common coding and design errors per coding standards. The Team makes an agreement to conform to these coding standards; thus, code analysis serves an integrity check with how well the Team is working according to their standards.
Code Coverage	A metric used to describe the degree to which source code has been tested. Code coverage is expressed as a percentage of lines of code tested over the total lines of code.
Commonality / Variability Analysis (CVA)	A technique that enables developers to write code that can be easily modified later. It complements iterative development. It is described in Shalloway and Trott's *Design Patterns Explained: A New Perspective of Object-Oriented Design*.
Constraint	Anything that limits a system from achieving higher performance or throughput; also, the bottleneck that most severely limits the organization's ability to achieve higher performance relative to its purpose or goal.
Consultant	An individual who has experience and expertise in applying tools and techniques to resolve process problems and who can advise and facilitate an organization's improvement efforts.

Term	Description
Continuous Process Improvement	A philosophy and attitude for analyzing capabilities and processes and improving them repeatedly to achieve customer satisfaction. Also known as Continuous Quality Improvement.
Cost-Benefit Analysis	An examination of the relationship between the monetary cost of implementing an improvement and the monetary value of the benefits achieved by the improvement, both within the same time period.
Customer	The recipient of the output (product, service, information) of a process. Customers may be internal or external to the organization. The customer may be one person, a department, or a large group. Internal customers (outside of IT) are sometimes called the "Business."
Customer Satisfaction	The result of delivering a product, service, or information that meets customer requirements.
Cycle Time	The total elapsed time to move a unit of work from the beginning to the end of a process. Cycle time includes process time, during which a unit is acted upon to bring it closer to an output, and delay time, during which a unit of work is spent waiting to take the next action.
Daily Stand-up	A stand-up meeting of the team where status is exchanged, progress is observed, and impediments are noted and removed. Typically, these meetings last 15 minutes. Each member answers three questions: • What did you do since last meeting? • What do you plan to do before next meeting? • What is blocking or slowing you down?
Dashboard	A type of information radiator that provides management and teams with graphs and reports indicating progress and trends and to identify potential problems.
Demonstrable Product	A version of a product that can be demonstrated to customers. It may not be quite ready for release or delivery, since that usually requires additional work (art work, production plans, etc). It is the primary deliverable of an iteration.

Lean-Agile Glossary

Term	Description
Design Pattern	An incomplete label for a collection of best practices for solving problems in a recurring context. The more general term is "Pattern" because patterns are involved with analysis, design, implementation, and testing.
Developer	Those members of the team who apply technical knowledge and skills to create the product. Testers may also be developers.
Done	The criteria for accepting a feature as finished. Specifying these criteria is the responsibility of the entire team, including the business.
Elevation	A planning tool that helps the team plan work iterations that minimize risk, increase learning, and coordinate effectively with other teams. Given all of the work required to realize an MBI, the team defines an objective for a particular iteration or set of iterations. The elevation is what is required to achieve that objective and ensure that the resulting code integrates well and maintains technical integrity.
Emergent Design	Allowing a design to emerge over time, as part of the natural evolution of a system. Requires good practices and testing to ensure that the system is not inadvertently allowed to decay or become overly complex over time. See *Emergent Design: The Evolutionary Nature of Professional Software Development* by Scott L. Bain.
Error Proofing	The ability to catch errors immediately, and to take corrective action to prevent problems down the line. Net Objectives identifies four strategies for error proofing: 1) eliminate possibility of error, 2) reduce occurrence if elimination is not possible, 3) reduce consequences of errors, 4) capture and address error early if they cannot be eliminated.
Facilitator	An individual specifically trained to enable groups to work more effectively, collaborate, and achieve synergy. The facilitator is a 'content neutral' party; does not take sides or express or advocate a point of view during the meeting, advocates for fair, open, and inclusive procedures to accomplish the group's work.

Term	Description
Feature	A feature is a business function that the product carries out. Features are large and chunky and are implemented by using many stories. Features may be functional or non-functional. The basis for organizing stories.
FIT	Framework for Integrated Test
Functional Test	The activity that validates a feature against customer requirements. They are usually done by a tester as part of the customer team.
Gemba	The "Gemba" is the place where value-added work is actually being done: a work cell, the developer team room, the help desk, the customer's office. Management must go to these locations to observe, evaluate, coach, and engage with the team. This is in contrast to management practices that rely on management hierarchies, team leads or formal status meetings in conference rooms or management offices.
Happy Path	The basic course of action through a single use case.
Harness Tests	A group of test scenarios with expected outcomes related to a specific system module to confirm that defects have not been introduced as a result of current iteration programming activity. A pricing test harness would confirm no defects from current pricing module programming.
Impediment	A technical, personal, or organizational issue that is preventing or delay in progress on delivering product.
Information Radiator	A type of visual control that displays information in a place where passersby can see it and get information about the project without having to ask questions. To be effective, the information must be current and must be easy to see and understand, with sufficient detail to explain status.
Integration System Test	The activity that verifies that software code does not harm other parts of the software product. Integration system test is usually done by a tester with automated tools.
Inventory	Work that has been started but not completed.

Lean-Agile Glossary

Term	Description
Iteration	A time-boxed period during which the team is focused on producing a demonstrable product, some amount of functionality that is ready to be shown to the customer and potentially ready to be delivered. Usually, iterations are 2-4 weeks long. In Scrum, an iteration is called a "sprint."
Iteration Plan	The list of user stories for the upcoming iteration.
Iteration Planning	The activity to prioritize and identify the stories and concrete tasks for the next iteration. Also known as "loading the front burner."
Just-in-Time-Design	The process of waiting until you know what the design needs to be and then refactoring code to meet these new needs before adding the functionality that is forcing the design change.
Kaizen	A Japanese term that means gradual unending improvement by doing little things better and setting and achieving increasingly higher standards. Masaaki Imai made the term famous in his book, *Kaizen: The Key to Japan's Competitive Success*.
Lean	The approach that produces value for customers quickly through a focus on reducing delays which results in increased quality and lower cost.
Lean-Agile	An approach to software development that incorporates principles, practices, and methods from lean product development, Agile software development, design patterns, test-driven development, and Agile analysis. Net Objectives advocates by this approach for those who want to be effective in creating products that add value to customers and to the business.
Manual Test	A document that lists the steps that a person follows to complete a test pass. Not automated.
Minimal Business Increment	The smallest set of functionality that must be realized in order for the customer to perceive value.

Term	Description
Open-Closed Principle	Suggested by Ivar Jacobsen, refined by Bertrand Meyer, and promoted by patterns, the Open-Closed Principle suggests that it is better to create designs that can accommodate change by adding to them, rather than by changing them. "Open-Closed" means we are "open to extension but closed to modification." This is a principle, and it is impossible to follow it literally at all times, but it can guide us in refactoring as well.
Pattern	A collection of best practices for solving problems in a recurring context, represented as collections of forces and provide a professional language for high-fidelity communication among developers. The subject of many books including *Design Patterns Explained* (Shalloway and Trott 2004).
PDCA	Plan-Do-Check-Act (PDCA) cycle is an iterative four-step problem-solving process for quality improvement. Also known as the Deming Cycle, Shewhart Cycle and Deming Wheel. • *Plan*: Development of plan to effect improvement. • *Do*: Plan is carried out. • *Check*: Effects of plan are observed. • *Act*: Results studied and learning identified for future usage.
Performance Test	A test that ensures the required level of performance of a product is met. This test checks both that the functionality works and that the time required to do the work is acceptable.
Persona	A description of the typical skills, abilities, needs, tasks, behaviors, and backgrounds of a particular set of users. As an aggregation, the persona is a fiction but is used to ensure groups of users are accounted for.
Planning	The activity that seeks to prioritize and define the stories and tasks for the next iteration. Also known as "loading the product backlog."
Process	A series of actions, changes, or functions performed in the making or treatment of a product.

Lean-Agile Glossary

Term	Description
Product	A collection of tangible and intangible features that are integrated and packaged into releases that offer value to a customer or to a market.
Product Vision	A short statement of the vision that is driving the project, expressed in business and customer terms: Who it is for, the opportunity, its name, the key benefit and differentiators. The Business Product Owner provides the product vision. Sometimes called the "Project Charter."
Project	A collection of releases, iterations, team members, and stories that creates a product. May have defined end dates or be on-going. May be: • All of the software a team or related teams are working on • A specific product or product group • A version of a company's product suite • A specific client implementation
Quality Assurance	A role and activity that ensures integrity, does release testing. The job of QA is to prevent defects from happening in the first place... it is *not* merely to find bugs.
Refactor	The disciplined technique for restructuring an existing body of code, altering its internal structure without changing its external behavior. Refactoring encourages and supports incremental design and implementation, the conversion of proofs of concept and early code into more suitable, stable code
Refactoring to the Open-Closed	The same tools and techniques that are used to clean up poor design and code (aka Legacy Refactoring) can be used to change a design just enough to allow for a clean addition or change to it. We take a design that was adequate initially, but which cannot now be changed cleanly to accommodate a new or changed requirement. Refactor until the code follows the open-closed principle, and then we integrate the new code.

135

Term	Description
Release	A version of a product that is released externally to customers. Releases represent the rhythm of the business and should align with defined business cycles. A release contains a combination of Minimal Business Increment that form a releasable product. A release may be internal and may be used for further testing. Two types of refactoring are fixing bad code and injecting better design into good code as requirements change.
Release Testing	The activity that validates that the product is good enough to release to customers. Release testing is typically performed by a tester and is sometimes called "Quality Assurance" (QA). Release Testing explores the statement, "We have built something that users can use." Often, release testing exposes requirements that fail to satisfy actual users' needs. See also: Quality assurance, Test.
Retrospection	Retrospection is the structured reflective practice to learn and improve based on what has already been done. The purpose of retrospection is to build team commitment and to transfer knowledge to the next iteration and to other teams. Retrospections must be done at the end of every iteration. A brief version, the "After Action Review" can be done at any time, whenever there is value for the team to stop and learn based on what has been done. In fact, the AAR can be more valuable because it allows the team to change while it can still help their current work.

Lean-Agile Glossary

Term	Description
Role	The role assumed by an individual on a given project. People may serve different roles on different projects or even different iterations. Scrum defines three roles: *The Business Product Owner.* Represents all customers and manages the Product Backlog. *The Team.* A combination of customers and developers and testers. The team is self-organizing within the guidelines, standards, and conventions of the organization. The team is responsible for completing the work in iterations. *The Scrum Master.* The Scrum Master is a facilitator for the team, responsible for coaching and ensuring that the Scrum process is used correctly, for helping to remove impediments.
Root Cause	A factor that caused nonconformance to plan and should be permanently eliminated through process improvement. QA helps lead RCAs.
Scrum	Scrum is an Agile process or framework for managing Agile projects. It is a project management process more than a methodology (the latter is rather too heavy). The Scrum Alliance is a group of analysts that originally developed the Scrum processes.
Scrum Master	Responsible for the process and the health of the team. Ensures that the team is fully functional and productive. Enables close cooperation across all roles and functions and removes barriers. Ensures that the process is followed. Facilitates the daily scrum, iteration reviews, and planning meetings. Also spelled "ScrumMaster."
Scrum Team (Team)	A cross-functional group comprised of persons with skills who can perform three roles – customer (requirements & validation), developers, and test. The team selects the iteration goal and specifies work results, has the right to do everything within the boundaries of the project guidelines to reach the iteration goal, and organizes itself and its work. Also known as the Team, an Agile team, or a work cell.

Term	Description
SIPOC	A diagramming tool used by Six Sigma process improvement teams to identify all relevant elements (suppliers, inputs, process, outputs, customers) of a process improvement project before work begins.
Story	A requirement, feature, and/or unit of business value that can be estimated and tested. Stories describe work that must be done to create and deliver a feature for a product. Stories are the *basic unit* of communication, planning, and negotiation between the Scrum Team, Business Owners, and the Business Product Owner.
Story Point	Story points are used in the Team Estimation game. They express relative complexity for purposes of estimating how much to place into an iteration. A common approach is to use Fibonacci numbers (1,2,3,5,8, ...) where 1 is low complexity, 8 is very complex) and then use Team Estimation or Planning Poker to get the team to agree on estimates.
Story Point Burn-up	The rate of progress the team is achieving in completing the project. It is measured as the number of story points completed per iteration and tracked toward the full project scope. Along with *velocity*, it visibly shows the duration and length of the current project.
Subject Matter Expert (SME)	A person who can speak with authority on an aspect of a project or who knows to whom to talk in order to get answers. Subject Matter Experts may represent a technology, the business, the customer, process, or any other topic of importance to the project.

Lean-Agile Glossary

Term	Description
Swarm	The activity of a teamlet that forms to complete a work item. The rules for swarming are:
	Focus on one story at a time. Within an iteration, teams should have only a few stories open at a time because the focus is on burning down stories. Do not dissipate energy by focusing on too much at once.
	The swarm is the priority. While individuals may work on other tasks, their priority should be the swarmed story.
	Swarms are Skill-based. If you have the skills to contribute to burning down a story, and you have capacity, you are expected to join in the teamlet, even if it is not exactly in your job description to do so.
Task	Tasks are descriptions of the actual work that an individual (or sub-team) does in order to complete a story. Typically, there are several tasks per story. Tasks have the following attributes:
	A *description* of the work to be performed, in either technical or business terms
	An *estimate* of how much time the work will take (hours, days)
	Exit criteria and *verification* method (test or inspection), and an indication of who will be responsible for the verification. All tasks must be verified, not just "done"
Teamlet	A group or sub-group of a Scrum team who agree to swarm on a work item (or story or task) in order to complete the work. The teamlet must have all of the knowledge and skills required to perform the work. Teamlets are generally formed at the Daily Stand-up, when the team reviews progress that has been made, and decide what needs to be worked on next. Teamlets are fluid, forming and disbanding based on the requirements of the work items at hand. The teamlet may assign a "Story Captain" to help the teamlet stay on track.

Term	Description
Test	Automatic and manual inspections of code and process to ensure correctness and completeness. Types of tests include Unit, Integration, System, Regression, Performance, and User Acceptance (Customer Acceptance)
Test-Driven Development	An evolutionary approach to development. In TDD, each test is written before the functional code that makes the test pass. The goal of TDD is specification and not validation, to think through a design before code is written, to create clean code that always works.
Tester	Those members of the team who apply testing knowledge and skills to validate and verify the product. Testers may be developers or customers and will use a combination of manual and automated methods. Testers may also act as consultants to developers to develop testing strategies.
Unit Test	The activity that verifies that software code matches the design specifications. Typically, unit testing is performed by a developer – by the code creator or by a "buddy" – however, testers often advise developers in testing approaches. Each unit test confirms that the code accurately reflects one intention of the system.
Use Case	In Lean-Agile, use cases express the details for a requirement story. If the use case becomes so large that it cannot be implemented in a single iteration, then the requirement story associated with the use case can be broken down into iteration requirement stories first. Alternatively, if a use case is more abstract, it can represent several requirement stories. Use cases should be expressed in the style of Cockburn detailed use cases. They may be white box or black box and include a main scenario, exceptions, and alternatives. Use cases can refer to business rules and data definitions.

Lean-Agile Glossary

Term	Description
User Acceptance Test (UAT)	The activity that verifies that software code matches the business intent. UAT belongs to the customer. They decide what constitutes an acceptable product. Unit tests help ensure that the acceptance tests are not about functional failures, but about the actual acceptability of the approach. Thus, UATs should not result in "it crashed" but «I would like the menus to be more descriptive."
Validation	The testing activity that confirms, "you built what I asked for."
Value-Added	A term used to describe activities that transform input into a customer (internal or external) usable output. An activity on a project is value-added if it *transforms the deliverables* of the project in such a way that the customer *recognizes* the transformation and is *willing to pay for it.*
Value Stream	The set of actions that take place to add value to a customer from the initial request to delivery. The value stream begins with the initial concept, moves through various stages for one or more development teams (where Agile methods begin), and on through final delivery and support.
Value Stream Owner	A person responsible for systematic management approach with immediate impact on the critical elements of a company's value streams. Makes change happen across departmental and functional boundaries.
Velocity	Velocity measures how many points the team spends during an iteration. The following velocities are used in the planning game to determine how many stories the customer should give to the team to estimate for the iteration: *Story velocity*. How many story points the team completes in an iteration. *Task velocity*. How many "engineering hours" the team actually can perform in an iteration.
Verification	The testing activity that confirms, "I built what I intended to."

Term	Description
Voice of Customer	The "voice of the customer" is the term used to describe the stated and unstated needs or requirements of the customer. The voice of the customer can be captured in a variety of ways: Direct discussion or interviews, surveys, focus groups, customer specifications, observation, warranty data, field reports, complaint logs, etc.
Waste	Any activity that consumes resources, produces no added value to the product or service a customer receives, or delays development of value. Types of waste include 1) anything that does not add customer value, 2) anything that has been started but is not in production, 3) anything that delays development, 4) extra features, 5) making the wrong thing.
Work Breakdown Structure (WBS)	The Work Breakdown Structure is "a deliverable-oriented grouping of project elements which organizes and defines the total scope of the project." The WBS organizes work according to three primary groups: • Product work or activities • Environment and team work or activities • Business work or activities
Work Cell	A cross-functional set of resources and people required to perform work in a value stream. In Lean-Agile, the preferred term for a work cell is "teamlet" because they tend to be temporary, forming to complete a particular work item.
XP (eXtreme Programming)	A software development methodology adhering to a very iterative and incremental approach. XP consists of a number of integrated practices for developers and management; the original twelve practices of XP include Small Releases, Acceptance Tests, On-site Customer, Sustainable Pace, Simple Design, Continuous Integration, Unit Tests, Coding Conventions, Refactoring Mercilessly, Test-Driven Development, System Metaphor, Collective Code Ownership, and Pair Programming.

Lean-Agile Glossary

Term	Description
Yesterday's Weather	The expectation that a team will complete as many story points worth of work in the next iteration as they did in the last. Of course, this will only be true after they have done a few iterations and have hit a somewhat steady level. Typically this is after 3-4 iterations. The term comes from the fact that before weather satellites, the most accurate way to predict weather was to say it would be the same as the day before. Hence, "yesterday's weather" means we expect what happened before.

Recommended Resources

Net Objectives Resources Page

www.netobjectives.com/resources

The main resources page for Net Objectives with sections for Lean, Agile, Scrum, Design, Testing, and Programming.

Net Objectives Scaled Agile Framework Resources

www.netobjectives.com/safe

Important resources for those wanting to use the Scaled Agile Framework (SAFe) to guide their Scrum practice. Net Objectives is the leading authority on SAFe outside of Scaled Agile. We offer over 13 SAFe related courses including: SPC, Architecture, shared services, Kanban, Product Portfolio Management. Three of our senior consultants have led large scale transitions at Fortune 50 organizations.

Net Objectives Training for Teams and Roles

www.netobjectives.com/training/scrum-and-kanban

Net Objectives has the most comprehensive and effective approach to preparing teams for the transition to Agile methods. While most Agile training companies focus on the Scrum Master, we focus on the enterprise. Our enterprise training enables your teams to transition to effectiveness. We also offer Lean-Agile Project Manager Certification courses, Lean Scrum Master Certification when appropriate.

Net Objectives Agile Design and Patterns Resources

www.netobjectives.com/resources/agile-design-and-patterns

Resources focused on the design, testing, and programming needs of developers. It has a wealth of materials on design patterns, agile design, object-oriented analysis and design.

Essential Skills for the Agile Developer

www.netobjectives.com/resources/books/essential-skills-agile-developers

Resources for the book, *Essential Skills for the Agile Developer.*

Recommended Resources

Acceptance Test-Driven Development Resources

www.netobjectives.com/atdd

Acceptance Test Driven Development (ATDD) is about communication, collaboration and clarity. This site offers resources help product owners and developers to use ATDD to gain clarity on requirements

Recommended Web Sites

Here are a number of very helpful web sites.

- Agile Alliance: *www.agilealliance.org*
- Agile Data: *www.agiledata.org*
- Agile Modeling: *www.agilemodeling.com*
- Code Qualities and the Open-Closed principle: *www.objectmentor.com/resources/articles/ocp.pdf*
- Crystal Methodology: *www.crystalmethodologies.org*
- Extreme Programming: *www.extremeprogramming.com*
- Gemba Panta Rei: *www.gembapantarei.com*
- Jim Highsmith: *www.jimhighsmith.com*

Recommended Books

A complete annotated bibliography may be found at *www.netobjectives.com/resources/bibliography*.

Bain, Scott L. *Emergent Design: The Evolutionary Nature of Professional Software Development*. Upper Saddle River, NJ: Addison-Wesley Professional, 2008.

Beck, Kent, and Cynthia Andres. *Extreme Programming Explained: Embrace Change* (2nd Edition). Boston, MA: Addison-Wesley Professional, 2004.

Bungay, Stephen. *The Art of Action: How Leaders Close the Gaps between Plans, Actions, and Results*. Boston, MA: Nicholas Brealey Publishing, 2010.

Collison, Chris, and Geoff Parcell. *Learning to Fly: Practical knowledge management from some of the world's leading learning organizations*. Chichester, West Sussex: Capstone, 2004.

VI Resources

Guernsey, Max III. *Test-Driven Database Development: Unlocking Agility.* Boston, MA: Addison-Wesley Professional. 2013.

Highsmith, James. *Agile Software Development Ecosystems.* Boston, MA: Addison-Wesley Professional, 2002.

Larman, Craig. *Agile and Iterative Development: A Manager's Guide.* Boston, MA: Addison-Wesley Professional, 2003.

Mann, David. *Creating a Lean Culture: Tools to Sustain Lean Conversions.* New York: Productivity Press, 2005.

Meszaros, Gerard. *xUnit Test Patterns: Refactoring Test Code.* Upper Saddle River, NJ: Addison-Wesley Signature Series, 2007.

Pugh, Ken. *Lean-Agile Acceptance Test-Driven Development: Better Software Through Collaboration.* Boston, MA: Addison-Wesley Professional, 2011.

Reinertsen, Donald G. *The Principles of Product Development Flow: Second Generation Lean Product Development.* Redondo Beach, CA: Celeritas Publishing, 2009.

Scholtes, Peter R. *The Leader's Handbook: Making Things Happen, Getting Things Done.* New York: McGraw-Hill, 1998.

Shalloway, Alan. "Lean Anti-Patterns and What to Do About Them." *Agile Journal.* 2008. *www.agilejournal.com/content/view/553/39/* (accessed 02 02, 2009).

Shalloway, Alan, and James R Trott. *Design Patterns Explained: A New Perspective on Object-Oriented Design,* Second Edition. Boston, MA: Addison-Wesley Professional, 2004.

Shalloway, Alan, Guy Beaver, and James R Trott. *Lean-Agile Software Development: Achieving Enterprise Agility.* Boston, MA: Addison-Wesley, 2009.

Shalloway, Alan, Scott Bain, Ken Pugh, and Amir Kolsky. *Essential Skills for the Agile Developer: A Guide to Better Programming and Design.* Boston, MA: Addison-Wesley Professional, 2011.

Sheridan, Richard. *Joy, Inc: How We Built a Workplace People Love.* New York: Portfolio Hardcover. 2013.

Sutton, James, and Peter Middleton. *Lean Software Strategies: Proven Techniques for Managers and Developers.* New York: Productivity Press, 2005.

Townsend, Patrick L, and Joan E Gebhardt. *How Organizations Learn: Investigate, Identify, Institutionalize.* Milwaukee, WI: ASQ Quality Press, 2007.

Womack, James P, and Daniel T Jones. *Lean Thinking: Banish Waste and Create Wealth in Your Corporation.* New York: Simon & Schuster, 1996.

VI Resources

Books from Net Objectives

Net Objectives consultants have written a variety of books and resources for business-driven software development. You can find them and other helpful resources at *www.netobjectives.com/marketplace*.

In the approximately ten years since the publication of the seminal work in the field of design patterns, this practice has moved from being an esoteric part of computer science research to the mainstream of software engineering. Yet despite their widespread acceptance, design patterns are frequently misunderstood.

Design Patterns Explained, Second Edition provides the reader with a gentle yet thorough introduction to design patterns and recent trends and developments in software design.

Scott L. Bain integrates the best of today's most important development disciplines into a unified, streamlined, realistic, and fully actionable approach to developing software. Drawing on patterns, refactoring, and test-driven development, Bain offers a blueprint for moving efficiently through the entire software life cycle, smoothly managing change, and consistently delivering systems that are robust, reliable, and cost-effective.

Essential Skills for the Agile Developer: A Guide to Better Programming and Design answers the question many developers have after taking some initial Agile/Scrum training – "OK, how do I write code now that we are building our software in iterations?" This book provides over a dozen proven practices that help developers improve their coding practices and make their code more easily changeable and maintainable in Agile projects.

For software to consistently deliver promised results, software development must mature into a true profession. *Emergent Design* points the way. As software continues to evolve and mature, software development processes become more complicated, relying on a variety of methodologies and approaches. This book illuminates the path to building the next generation of software.

Software development projects have been adopting agility at a rapid pace. Although agility provides quicker delivery of business value, lean principles suggest reducing waste, delays, and hand-offs can provide even faster delivery. With *Lean-Agile Acceptance Test-Driven Development: Better Software Through Collaboration* to help, the business customer, the tester, and the developer collaborate to produce testable requirements.

These acceptance tests form the detailed specification of how the software should work from an external point of view. They help the customer to clarify their needs, the developer to have an objective to code towards, and the tester to plan for more than just functional testing.

The *Lean-Agile Pocket Guide for Scrum Teams* is a useful reference for Scrum teams who have had some basic training and want to use Scrum in the context of Lean. Topics include the essential competencies of Scrum, how to get started, how Lean and Scrum relate, the roles of Scrum, planning and analysis and estimation, iterations, quality and testing, and communication. This pocket guide offers a complete set of checklists for Scrum teams and resources teams need to succeed.

Lean-Agile Software Development: Achieving Enterprise Agility addresses how to drive maximum value from Lean development and avoid or fix the mistakes that prevent software teams from succeeding with Lean, the crucial make-or-break details that team leaders and developers need to succeed with Lean processes, and why many teams fall back on ineffective processes that compromise their commitment to Lean software development.

VI Resources

More and more software organizations are recognizing the potential value of Lean techniques for improving productivity and driving more business value from software. It requires clarity, knowledge and skills that many organizations haven't developed.

This book brings together practical insights every organization needs to succeed with Lean. The authors answer the most important questions about Lean development: What tools can I use to successfully implement Lean in my company? How do I transition to Lean Software Development? How do I correct specific counterproductive practices that stand in my way? How do I identify waste within my company?"

Lean Software Strategies: Proven Techniques for Managers and Developers shows how the most advanced concepts of lean production can be applied to software development and how current software development practices are inadequate.

Lean production, which has radically benefited traditional manufacturing, can greatly improve the software industry with similar methods and results. This transformation is possible because the same over-arching principles that apply in other industries work equally well in software development. The software industry follows the same industrial concepts of production as those applied in manufacturing; however, the software industry perceives itself as being fundamentally different and has largely ignored what other industries have gained through the application of lean techniques.

Written for software engineers, developers, and leaders who need help creating lean software processes and executing genuinely lean projects, this book draws on the personal experiences of the two authors as well as research on various software companies applying lean production to software development programs.

Prefactoring approaches software development of new systems using lessons learned from many developers over the years. It is a compendium of ideas gained from retrospectives on what went right and what went wrong in development. Some of these ideas came from experience in refactoring. Refactoring is improving the design of existing code to make it simpler and easier to maintain.

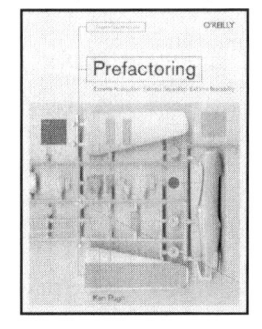

This practical, thought-provoking guide details prefactoring guidelines in design, code, and testing. These guidelines can help you create more readable and maintainable code in your next project. To help communicate the many facets of this approach, Prefactoring follows the development of a software system for a fictitious client, named Sam, from vision through implementation.

In *Test-Driven Database Development*, Max Guernsey, III shows how to adapt Test-Driven Development (TDD) to achieve the same powerful benefits in database design and development.

Guernsey explains why TDD offers so much potential to database practitioners and how to overcome obstacles such as the lack of conventional "testable classes." He shows how to use "classes of databases" to manage change more effectively; how to define testable database behaviors; how to maximize long-term maintainability by limiting a database's current scope; and how to use "emergent design" to simplify future expansion.

This book is a guide to applying the proven practice of TDD to database needs: organizing and optimizing the organization's data for a significant competitive advantage.

VI Resources

Ordering and Licensing Information

This pocket guide is available in a volume discount from the Net Objectives Marketplace. For pricing, visit *www.netobjectives.com/marketplace*.

License (Private Label) this Pocket Guide for Your Company

Many companies have found it useful to adapt this pocket guide to support their own Agile methodology. We can work with you to create a print version or to embed it in your web-based tool.

Please contact *info@netobjectives.com*.

Index

A

Acceptance TDD 13; Resources 144

Agile; Essential principles 11; Flow of value 15; Life-cycle 11; Rationale 10; Rewards 6; Risks 6; Starting an Agile team 7

Analysis 54

Application Development Manager (ADM); and Technology Delivery Manager 24; Defined 24

B

Business Analyst; Defined 32; Standard work 33

Business case for Scrum 3

Business delivery 45, 115, 117, 120; Stages and outcomes 46

Business discovery 45, 113, 114; Activities 49; Stages and outcomes 46

Business Product Owner 4, 10, 38, 42, 48, 51; Defined 27; Standard work 27, 28

Business SME. *See* Business Analyst

C

Capacity 5

Checklists; Business delivery 115, 117, 120; Business discovery 113, 114; Daily stand-up 122; Iteration 0 62, 115; Iteration execution 117; Iteration planning 117; Planning 60; Priority 113; Product demonstration 71, 124; Ready-to-Pull 114; Release 120; Release planning 112, 113; Staging 114; Starting an Agile team 7, 111

Co-location 4, 5, 50, 110

Commonality / Variability Analysis (CVA) 129

Communication 42, 48, 79, 107

Continuous improvement 9, 11, 13, 14, 16, 20, 130

D

Daily Scrum. *See* Daily Stand-up

Daily Stand-up 3, 10, 15, 28, 29, 67, 130
Daily work 28, 31, 33, 35, 37, 39, 40, 41
Database Administrator; Defined 40; Standard work 40
Database Developer; Defined 41; Standard work 41
Demonstration 70
Design Patterns. *See* Patterns, Design; Resources 144
Design Patterns Explained (book) 129, 148
Developer; Daily work 35; Defined 34

E

Eisenhower 45, 88
Elevation 53, 57, 59, 131
Emergent Design (book) 148
Essential competencies of Scrum 3
Estimation 50; Team Estimation Game 81

F

Fast-Flexible-Flow 12, 13
Feature 28, 34, 49, 76, 81
Flow of value 15

G

Gemba 15
Glossary 127
Ground rules; Daily stand-up 122; Product demonstration 120

I

Impediments 28, 38, 88, 130, 132
Implementing Lean-Agile for your team by Net Objectives 1
Information radiator. *See* Visual control
Iteration 60; Iteration execution 64; Iteration planning 62
Iteration 0 61

J

Just-In-Time 13, 18, 133

K

Kanban. *See also* Visual control

Knowledge 6, 10, 13, 16, 83, 84

L

Lean-Agile; Foundations 15; Practices 18; Principles 17

Lean software development 16; Features of Lean 9; Practices 18; Principles 9, 17

M

Metrics 10, 13, 99, 111

Minimal Business Increment (MBI) 3, 38, 55, 56, 74, 133, 136

Minimum Marketable Feature (MMF). *See* Minimal Business Increment (MBI)

N

Net Objectives; Net Objectives marketplace 148; Net Objectives resources 144

P

Patterns, Design 10, 17, 20, 78, 83, 93, 131, 134, 148

Plan-Do-Check-Act (PDCA) 79

Planning 45; Thinking activities 52

Poppendieck 12

Principles 11, 12, 36, 83

Private label this pocket guide 152

Product team. *See* Team

Product vision 12, 51, 72, 135

Project board. *See* Visual control

Q

Quality 83; Code quality 94; Continuous build 15, 18; Continuous improvement 14, 16, 83; Continuous integration 99

Quality Assurance 4; Daily work 37; Defined 34, 35

R

Refactor 20, 135

Release Manager; Defined 38; Standard work 39

Retrospection 84

Right-sized 4

Risks 6

Roles 4, 21, 42, 137. *See also* specific roles; Essential skills 19, 144; Front-Line level 23; Management level 22; Portfolio level 22; Swarm 4, 43; Team 4

S

Scaled Agile Framework (SAFe); Resources 144

Schedule; Iteration Planning 125; Scrum of Business Product Owners 124; Scrum of Lead TDMs and Business Product Owners 124; Scrum of Scrum Masters 124; Scrum of Technology Delivery Managers 124

Scrum 1, 137; Business case for Scrum 3; Daily Stand-up 67; Daily work 28, 35, 39, 40, 41; Good Scrum practices 4; Risks and rewards 6; Roles defined 4, 32, 34, 37, 40

Scrum Master 4, 15, 29, 83, 137; Defined 29; Standard work 31

Size 3, 32, 74

Sprint. *See* Iteration

Sprint 0. *See* Iteration 0

Story 75, 138; Completing stories 66; Essential elements 3; Right sized 4

Story board 65

Sustainable 11

Swarm 4, 139; Defined 44

Index

T

Team 4, 10, 14, 15, 43; Defined 42

Teamlet; Defined 44

Team project board. *See* Visual control

Technology Delivery Manager (TDM); and Application Development Manager 24; Defined 25

Testing 10, 17, 35, 95, 140; Acceptance testing 95, 127; Automated testing 96; Roles in testing 66; Test-driven development (TDD) 140; Test environment 106

Training by Net Objectives; Implementing Scrum for Your Team 1; Kanban 144; Roles 144; Scaled Agile Framework (SAFe) 144; Scrum 144

V

Value-added 1, 3, 141

Value stream 10, 15, 17, 45, 52, 90, 141

Velocity 80

Visual control 10, 17, 66, 108. *See also* Gemba

NET OBJECTIVES LEAN-AGILE APPROACH

INTEGRATED AND COHESIVE

All of our trainers, consult-ants, and coaches follow a consistent Lean-Agile approach to sustainable product development. By providing services at all of these levels, we provide you teams and management with a consistent message.

PROCESS EXECUTION

Net Objectives helps you initiate Agile adoption across teams and management with process training and follow-on coaching to accelerate and ease the transition to Lean-Agile practices.

SKILLS & COMPETENCIES

Both technical and process skills and competencies are essential for effective Agile software development. Net Objectives provides your teams with the know-ledge and understanding required to build the right functionality in the right way to provide the greatest value and build a sustainable development environment.

ENTERPRISE STRATEGIES

Enterprise Agility requires a perspective of software development that embraces Lean principles as well as Agile methodologies. Our experienced consultants can help you develop a realistic strategy to leverage the benefits of Agile development within your organization.

We deliver unique solutions that lead to tangible improvements in software development for your business, organization and teams.

SERVICES OVERVIEW

TRAINING FOR AGILE DEVELOPERS AND MANAGERS

Net Objectives provides essential Lean-Agile technical and process training to organizations, teams and individuals through in-house course delivery worldwide and public course offerings across the US.

CURRICULA — CUSTOM COURSES AND PROGRAMS

Our Lean-Agile Core Curriculum provides the foundation for Agile Teams to succeed.

- Lean Software Development
- Implementing Lean-Agile for Your Team
- Agile Enterprise Release Planning
- Sustainable Test-Driven Development
- Agile Estimation with User Stories
- Design Patterns

In addition, we offer the most comprehensive technical and process training for Agile professionals in the industry as well as our own Certifications for Scrum Master.

PROCESS AND TECHNICAL TEAM COACHING

Our coaches facilitate your teams with their experience and wisdom by providing guidance, direction and motivation to quickly put their newly acquired competencies to work. Coaching ensures immediate knowledge transfer while working on your problem domain.

LEAN-AGILE ASSESSMENTS

Understand what Agility means to your organization and how best to implement your initiative by utilizing our Assessment services that include value mapping, strategic planning and execution. Our consultants will define an actionable plan that best fits your needs.

LEAN-AGILE CONSULTING

Seasoned Lean-Agile consultants provide you with an outside view to see what structural and cultural changes need to be made in order to create an organization that fosters effective Agile development that best serves your business and deliver value to your customers.

FREE INFORMATION

CONTACT US FOR A FREE CONSULTATION

Receive a free no-obligation consultation to discuss your needs, requirements and objectives. Learn about our courses, curricula, coaching and consulting services. We will arrange a free consultation with instructors or consultants most qualified to answer all your questions.

Call toll free at 1-888-LEAN-244 (1-888-532-6244) or email sales@netobjectives.com

REGISTER PROFESSIONAL LEAN-AGILE RESOURCES

Visit our website and register for access to professional Lean-Agile resources for management and developers. Enjoy access to webinars, podcasts, blogs, white papers, articles and more to help you become more Agile. Register at www.netobjectives.com/user/register.

Net Objectives

Contact Us:
sales@netobjectives.com
1-888-LEAN-244
(1-888-532-6244)